BATTLESHIPS

ANTONY PRESTON

BATTLESHIPS

ANTONY PRESTON

CRESCENT
A BISON BOOK

This edition is published by
Crescent Books,
a division of Crown Publishers, Inc.
One Park Avenue
New York
New York 10016

First English edition published by
Bison Books Limited
4 Cromwell Place
London SW7

Library of Congress Cataloging in Publication Data
Preston, Antony, 1938—Battleships.
    1. Battleships—History. I. Title
V815.P73   359.8'3

Library of Congress Catalog Card Number 80 25806
ISBN 0 517 33002 4
a b c d e f g h
Printed in Hong Kong by Toppan Printing Co., (H.K.) Ltd.

Reprinted 1982

CONTENTS

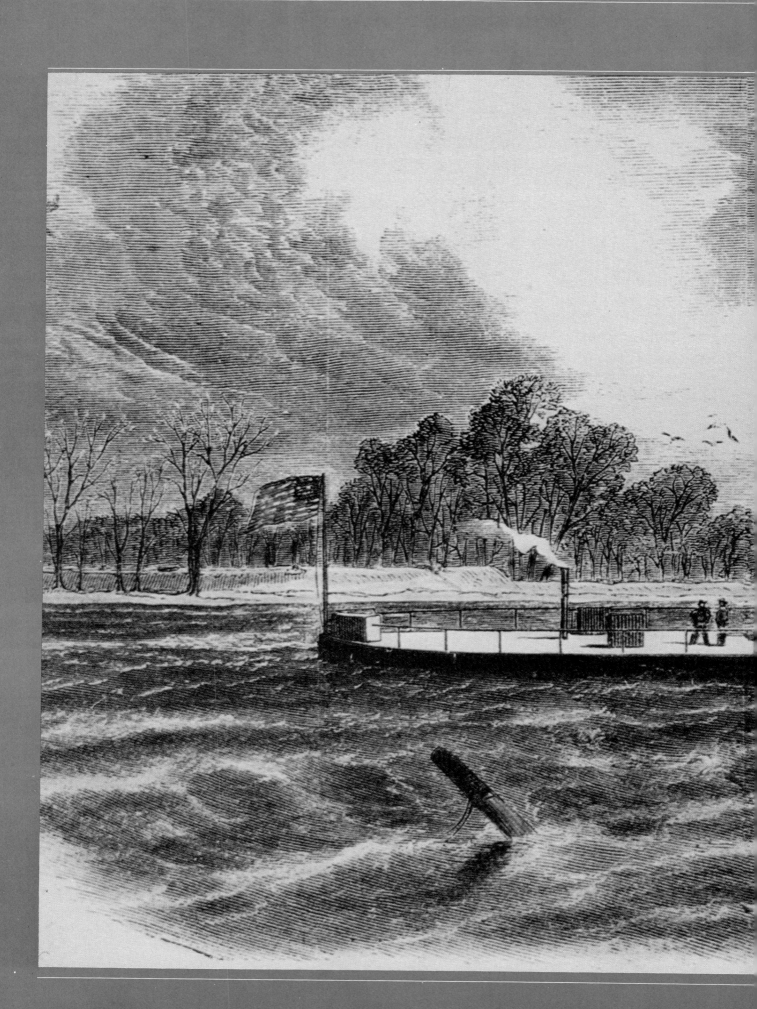

1. SHIP OF THE LINE

No ship has ever achieved an aura of power and might to match that of the battleship. Others have been more graceful, faster and even more powerful, but throughout their comparatively short reign battleships were publicly equated with material and political power. In days gone by national strength was reckoned in dreadnoughts and even as late as 1967, long after the battleship counted as a unit of strength, the US Marines were demanding that a battleship should be recommissioned to provide heavy-gun support in the Vietnam War.

To understand where it all started we have to go back five centuries, to the time when it became feasible to mount heavy guns on board ships. A Frenchman called Descharges is traditionally regarded as the inventor of a method of cutting gunports in the lower deck and fitting them with hinged lids, about 1500. The problem was not merely that of cutting a port, for loading ports had been seen in merchant ships for much longer, but how to avoid cutting into the main longitudinal strength of the ship, the timber 'wales.' At first gunports were provided only at the after end of the lower deck, to keep them well clear of the waterline but as the art of shipbuilding improved, the height of the lower deck could be raised and it was not long before a complete bank of guns was possible. There is an old English tradition that King Henry VIII was the first to pierce ships' sides for heavy guns, but whoever gave the first order it seems that the first English man-of-war with a complete lower gundeck appeared sometime after 1515 but well before 1546. This transition was also being made by the other navies of Europe and it marks a shift away from a style of sea warfare which had lasted since ancient times. Hitherto the basic pattern of fighting was with land-weapons between opposing groups of *men*, but from now on the emphasis would be on damaging the enemy *ship*. Boarding and hand-to-hand fighting would last for another four centuries but from the sixteenth century anti-ship weapons assumed ever more importance.

Nothing illustrates the transition better than a comparison between the Battle of Lepanto in 1571 and the various actions involving the Spanish Armada in 1588. Guns were used with deadly effect at Lepanto but with the primary objective of killing sufficient rowers to cripple an enemy galley or to cause sufficient casualties to prevent boarding attempts. Off Start Point

Previous page: Ericsson's 'steel-clad battery,' the USS *Monitor* as seen by the readers of *Harpers Weekly* in March 1862.

Below: The Four Days' Fight of 1–4 June 1666, a bloody melée between the Dutch and English in the Second Anglo-Dutch War.

on 31 July 1588 Howard of Effingham used the *Ark Royal*'s culverins to pound Juan Martinez de Recalde's galleon *San Juan de Portugal* at long range. In their reports the English captains talk admiringly of the number of short-range 'ship-smashers, cannon and periers' carried by the best Armada ships. Of course intention was not the same as achievement and neither side did much damage to the other in the early fighting; even the biggest Armada guns lacked the sort of penetrating power needed to make an impression on the heavy timbers of a big ship. At a range of 700 yards a shot from a culverin or demi-culverin might well easily fail to go through the hull and even if it did the hole could be plugged by the crew.

In the following century the lesson of the Armada was seen to be that ships could be sunk by gunfire, but only with concentrated fire at much shorter range. The logical outcome of this was greater discipline in tactics, marshalling the ships into a 'line of battle' and increasing the strength of the 'battery' or 'broadside.' What made these aims easier to achieve were radical developments in the science of naval architecture, and it can be said that most of the problems of sailing warship design were solved during that century. A more scientific approach to naval architecture was accompanied by a growing political awareness of sea power, and it is in the seventeenth century that the prestige ship becomes common. Louis XIV's Navy Minister, Colbert, wrote that nothing was more impressive or so befitted the majesty of the King as a ship bearing the finest decoration. Christian IV of Denmark had his *Tre Kroner*, Gustavus Adolphus of Sweden built the *Wasa*, England built the *Sovereign of the Seas* and France had the last word with the magnificent *La Couronne* in 1638. The lust for glory often outstripped any military needs, and royal folly and lack of taste culminated in the loss of the *Wasa* in 1628. She sank on her maiden voyage, capsized in Stockholm harbor by a sudden gust of wind, because she was insufficiently ballasted.

The need for more gunpower had already forced designers to add a second tier of guns, and by the early years of the century three-deckers were being built. Gun design was improving, and some of the bewildering nomenclature of the previous century had been weeded out. The 100-gun *Sovereign of the Seas* (1637) carried 20 cannon drakes and eight demi-cannon on the lower tier, 24 culverin drakes and six culverins on the middle tier, and 24 demi-culverin drakes, four demi-culverins, 16 demi-culverin drakes and two culverin drakes on the upper tier, waist, forecastle and

quarterdeck. A 'Drake' was shorter and lighter than an ordinary gun of the same caliber; cannon were later known as 42-pounders, demi-cannon as 32-pounders, culverins as 18-pounders and demi-culverins as 9-pounders. All but the smallest guns were now muzzle-loaders and the old method of building up guns from iron segments had given way to casting.

As the science of design stabilized it became possible to divide the major warships into six classes or 'rates,' the first two being the most important, including all the three-deckers, the third and fourth including various grades of two-deckers and the fifth and sixth covering ships down to 20 guns. But a ship rated as a 100-gunner did not necessarily carry 100 guns – the *Sovereign of the Seas*, for example, carried 116 guns without counting small anti-personnel weapons such as swivels. The rating system was to change constantly for another two hundred years, for it was a subtle formula which took into account the size of crew as well as the number of guns.

By the end of the century standardization had progressed further, with guns being designated by the weight of shot. Iron had virtually replaced bronze and brass, giving greater strength and longer life. So far had the design of the capital ship progressed that it can be said that in broad outline there was little

Above: The characteristics of the ship of the line were first seen in the galleons of the Armada period. This Elizabethan galleon model is based on the only surviving plans of the period.

difference in the design of, say, an English 100-gun ship of 1700 and one of 1800. This is not to say that no progress occurred, for in every other respect the eighteenth century saw many more developments in the design of rigging and ordnance. What was changing, too, was the way in which each navy approached the design of its warships. The Dutch, Swedes and Danes, for example, needed to keep down draught to navigate the shallow waters around their coasts and like the French favored lightly built, fast ships. The English, in contrast, remembered the importance of gunpower from the Armada days and insisted on heavy armament. They also had the problem of operating in less sheltered waters and had to have robust ships. Contemporary accounts tend to harp on their sluggishness but admit that in close-range action an English man-of-war could take more punishment.

Punishment is the right word for these naval actions. The Second Dutch War of 1665–67 saw a series of bloody encounters between the rival fleets in which large numbers of men and

ships were lost, such as the Four Days' Fight and the Battle of Scheveningen. The art of signalling improved, making it easier for ships to concentrate their fire on the part of the enemy fleet which the commander selected. In

Above: The quarterdeck of HMS *Victory* at the height of the Battle of Trafalgar. Nelson (foreground) has just been hit by a bullet.

Below: The Glorious First of June, 1794, with the two-decker HMS *Brunswick* successfully engaging the French *Vengeur du Peuple* (right) and *Achille* (left).

a word, the ship of the line was becoming an instrument of comparative precision, capable if soundly handled of doing great execution to her enemies.

During the eighteenth century the steady growth in size and fighting power went on. The 1st Rate of 100 or more guns was expensive to build and man, and relatively few could be afforded so a much greater number of three-decked 2nd Rates (88–98 guns) and two-decked 3rd Rates (64–84 guns) made up the balance of the fleet. The older two-deckers were no longer regarded as fit to 'lie in the line' but remained useful for escorting convoys, acting as flagships on distant stations and could even in an emergency stiffen the battle line.

These 4th Rates (50–60 guns) and 5th Rates (32–44 guns) were the eighteenth-century counterpart of the heavy cruiser in World War II.

The 74 became the standard battleship of the latter part of the eighteenth century, taking over from the 64 as the work-horse of the fighting fleets. At the Battle of the Nile the British had nothing larger than 74s, for example, and with their maneuverability and better turn of speed they provided a greater degree of tactical flexibility. A typical British 74 of 1793 would carry 28 32-pounders on the lower deck and 28 18-pounders on the upper deck as well as 14 12-pounders on the quarter-deck and four more on the forecastle. She might also carry a pair of carronades on the forecastle but these would not affect her rating as a 74-gun ship. These short-barrelled guns were the Royal Navy's secret weapon in the War of the American Revolution, firing a very large shot over a short distance but light enough to be handled by a small gun crew. Because of their deadly effect on ship timbers at close range they were first known as smashers but because they were cast at the Scottish Carron Foundry they were soon dubbed carronades.

The line of development which had started in the early 1500s reached its peak at Trafalgar in 1805, when the main British Fleet under Vice-Admiral Lord Nelson met and destroyed a superior Franco-Spanish Fleet under Admiral Villeneuve. Nelson had three 100-gun ships, his own *Victory*, Collingwood's *Royal Sovereign* and Northesk's *Britannia* as well as two 98s while Villeneuve had the giant 130-gun *Santisima Trinidad*, the 112-gun *Santa Anna* and *Principe de Asturias* and the 100-gun *Rayo*. But the battle would be decided by the 2nd and

3rd Rates, from the 98-gun *Dreadnought* and the 80-gun *Bucentaure* down to the elderly 64s *Agamemnon*, *Africa*, *Polyphemus* and *San Leandro*.

From noon until evening the battle raged, and at the end Nelson, although dying, knew that he had achieved his single ambition of a decisive battle of annihilation. His own fleet survived intact, although many ships had been severely damaged and 1700 men had been killed or wounded; but 19 French and Spanish prizes had been taken and some 20,000 prisoners. The French 74 *Achille* had been set on fire during the battle and had blown up, and no fewer than 17 French and Spanish ships had struck their colors. It was a victory without equal in history, and however much was due to training and seamanship, French and Spanish survivors' accounts harp on the high standard of the English gunnery. Small wonder that the Royal Navy in years to come talked of the 32-pounder as 'the gun which won the Great War against France.'

Although fighting at sea continued for another nine years Trafalgar made all else seem an anticlimax, and the main fleets did not meet again. But the culmination of nearly two decades of war was that the design of the sailing ship of the line was brought to a degree of perfection limited only by the materials and knowledge available. In 1808 the Royal Navy launched the 120-gun *Caledonia*, whose 205-foot gun deck was widely accepted by designers as being near the maximum practicable with a single wooden keel. Then in 1813 the Surveyor of the Navy, Sir Robert Seppings found a way to preserve longitudinal strength by using diagonal wooden trusses and iron stiffening to prevent excessive hogging and sagging. Thus to talk of wooden walls surviving to the middle of the nineteenth century is misleading, for the

later generation of three-deckers had a good deal of iron in their wooden hulls. Other improvements were made to the bow and stern, to avoid the fearful damage done to ships like HMS *Victory* at Trafalgar. A new round bow, with the heavy timbers taken up to the forecastle deck end, allowing guns to bear directly ahead, was made standard in 1811. A similar job of redesign was done on the stern, replacing the flimsy stern galleries with a strongly timbered 'circular' stern.

Into this cosy world of slow majestic progress irrupted an innovation that was ultimately to destroy it. In 1822 the French artilleryman Henri Paixhans published a treatise on how the French navy could avenge its recent shattering defeat. He argued that as France had no hope of building a new navy to replace the one destroyed in 1793–1815 she must find a technological answer, an 'equalizer' which would wipe out the British lead in warships. His solution was not a new one, merely an adaptation of an old idea; he suggested that the hollow cast-iron mortar bomb, which had been around since the seventeenth century, could be redesigned to be fired from the 'long' ship gun. If the shell was to lodge in an enemy ship's timbers it would tear an enormous hole and probably start a fire, and there was nothing that a contemporary sailor feared as much as fire, for with her mass of tarred cordage and dry timbers a wooden man-of-war was likely to be a deathtrap. For this reason 'infernal machines' and devices such as red-hot shot had found little favor; it was often recalled that Admiral Brueys' flagship *L'Orient* had exploded at the Battle of the Nile in 1798 because of some explosive contraption or other carried on board.

Paixhans had little trouble in getting his ideas accepted, but instead of a fleet of small steamers with shell guns demolishing the Royal Navy he had to endure the depressing sight of the British equipping their ships with his gun. Once both sides had the 'equalizer' they were back where they started. Because it was difficult to maintain the strength of a hollow cast-iron sphere the first shell guns were of large caliber, the French choosing the 55-pounder and the British the roughly comparable 68-pounder (as the livre was heavier than the British pound there was not much difference between the calibers). The French introduced their *canon-obusier* in 1824 and the British issued their shell guns two years later, but both navies continued to make solid-shot guns for the simple reason that they were more accurate over longer ranges. By the end of the 1830s most three-deckers carried a mix of some 60 percent of the standard type to 40

Below: This imaginary group of Nelson's flagships, *Agamemnon*, *Captain*, *Vanguard* and *Victory* shows the difference in bulk between two-deckers and three-deckers.

percent shell guns. The method of ignition was primitive but effective: the wooden fuse was ignited by the flash of the black powder charge as the gun fired, and a simple time-delay prevented detonation until (it was hoped) the shell struck the target.

Great Britain was still in the throes of her Industrial Revolution and it was not long before the steam engine was tried in ships. As early as 1814 the Royal Navy tried to propel a small sloop by steam and seven years later the first paddle sloops were ordered from the Royal Dockyards. There was no attempt to build a steam-powered battleship at this stage for one very simple reason: no new ships were laid down in the years after Waterloo because so many hulls were already on the stocks. In the late 1820s and early 1830s a new class of 120-gun 1st Rates was begun, the *Trafalgar*, *Prince Regent*, *Royal George*, *Neptune*, *Royal William*, *Waterloo* and *St George*. Their armament was to be six 68-pounders and 114 32-pounders, with a complement of 900 men, and yet their length of 205 feet was less than a minesweeper of World War II. Even bigger ships followed, for the Royal Navy had a huge investment in wooden shipbuilding. Nor was the political climate favorable to rapid change, given the contemporary philosophy which

preached that material progress would make war obsolete.

Expenditure on new battleships might be limited but the spirit of the age was an enquiring one, and it was only a matter of time before steam power was installed in a big warship. The paddle wheel was clearly only suited to small warships for it was vulnerable and would interfere with too large a percentage of the broadside guns. Following trials with the new screw propellor their Lordships decided as early as May 1840 to build a screw steamship, the sloop *Rattler*. In November 1845 work started on converting the incomplete 3rd Rate

Above: The two-decker HMS *London* was laid down as a sailing ship but was converted to steam in 1860 and lasted as a depot ship until 1884.

Below: The 130-gun *Bretagne*, launched in 1855, was one of the last French three-deckers. She became a training hulk in 1866.

Marine Militaire. — Le Navire-Ecole « La Bretagne »

Ajax into a screw 'guardship' and when she went to sea on 23 September 1846 she was the world's first seagoing steam battleship. As she and her sister *Edinburgh* carried only 58 guns they were theoretically inferior to other two-deckers but it was recognized that they could outmaneuver any sail-powered warship afloat and so they were given the vague designation of 'screw blockships,' on the understanding that they could perform blockade duties.

Thereafter the pace accelerated as both the French and British put steam engines into all new ships of the line and converted as many of the existing hulls as they could afford. By 1853, on the eve of the Crimean War the screw propeller and the shell gun were firmly established, although more perceptive observers were beginning to talk of the towering 131-gun three-deckers as 'eggshells armed with hammers,' and it was reckoned that a future naval engagement would last minutes rather than hours. The small actions of the 1840s had shown that shells worked, but it was not until November 1853, when a Russian squadron of six ships of the line and four smaller ships trapped and sank a Turkish squadron of 11 smaller ships at Sinope, that the major navies took the threat to their ships seriously. It would have been tactless to point out that Admiral Nakhimov took several hours to sink a force of 10 frigates and smaller warships trapped in the anchorage, just as Admiral Codrington's Anglo-French squadron had done to the Turks at Navarino in 1827. What

stamped Sinope as a massacre was the fact that all the Turkish ships had been set on fire, and what lent it even more sinister significance was the fact that the British and French, with their large fleets of wooden sailing ships, had just let themselves be drawn into a war with the victor of Sinope.

The Anglo-French fleet was to get a taste of the new technology when it tried to support the land attack on Sevastopol with a bombardment of the forts. On 17 October 1854 the lumbering two-deckers and three-deckers went into action, the sailing ships each being towed by a small steamer. After about an hour and a half the *Albion* had been set on fire twice and sustained severe casualties from four shell hits. She was towed out of action stern first, and then it was the turn of the *Queen* to be set on fire by red-hot shot and the frigate *Arethusa* to be knocked about. The French *Ville de Paris* also suffered from a mortar shell under the poop which caused many casualties, but apart from the *Albion* all damaged ships were ready for action next day. Despite this comparatively modest total of damage the bombardment of Sevastopol has been frequently described as a disastrous defeat for the Allies. The ships had not inflicted much damage on the Russian forts, which is hardly surprising as they were restricted by shallow water to a distance of 750 yards, when it was known that 500 yards was the maximum effective range.

Although the Allies had got off lightly they took energetic steps to make sure that the next

Below: The gaunt hull of the floating battery *Terror* ready for launching at Jarrow in April 1856, too late for the Crimean War.

time would be on more favorable terms. The French Navy's *Directeur du Matériel* immediately suggested filling the hollow sides of a ship with cannon balls, but the Royal Navy came up with a more sensible idea, 4-inch wrought-iron plating. Both countries had experimented with wrought-iron warship hulls in the 1840s but had lost interest when it became obvious that iron tended to splinter when struck by solid shot. But by 1854 the science of metallurgy had advanced and wrought-iron now had sufficient elasticity to be able to absorb the energy of a hit without shredding into lethal splinters.

The French and British collaborated on the hurried design of 10 fighting batteries, more like floating packing cases than ships of the line, with a single gun deck, 4-inch plating on the sides and a simple barque rig and a steam engine to drive them at 4 knots. With bow and stern so bluff as to be almost square, they could hardly steam or sail adequately, but they were intended to be towed to their theater of operations. The French ships, called *Congreve*, *Devastation*, *Foudroyante*, *Lavé* and *Tonnant*, were ordered shortly after the Sevastopol bombardment and were ready the following summer, and the British *Aetna*, *Glatton*, *Meteor*, *Thunder* and *Trusty* were ready in April 1855. A series of minor delays prevented the British floating batteries from getting out to the Black Sea in time for a big assault on the forts at the mouths of the Dnieper and Bug rivers, and so the honor for the first action with

armor-plated ships goes to the French. On 17 October 1855 three French batteries opened fire with their 55-pounder guns against Kinbourn Kosa, a group of five forts guarding the approaches to Odessa. An hour and a half later the forts surrendered, having seen that the ships were impervious to the rain of red-hot shot and shell. Despite being hit repeatedly the

Above: The French floating battery *La Foudroyante*, one of the original five launched in 1855, in dock with a sister behind, in 1869.

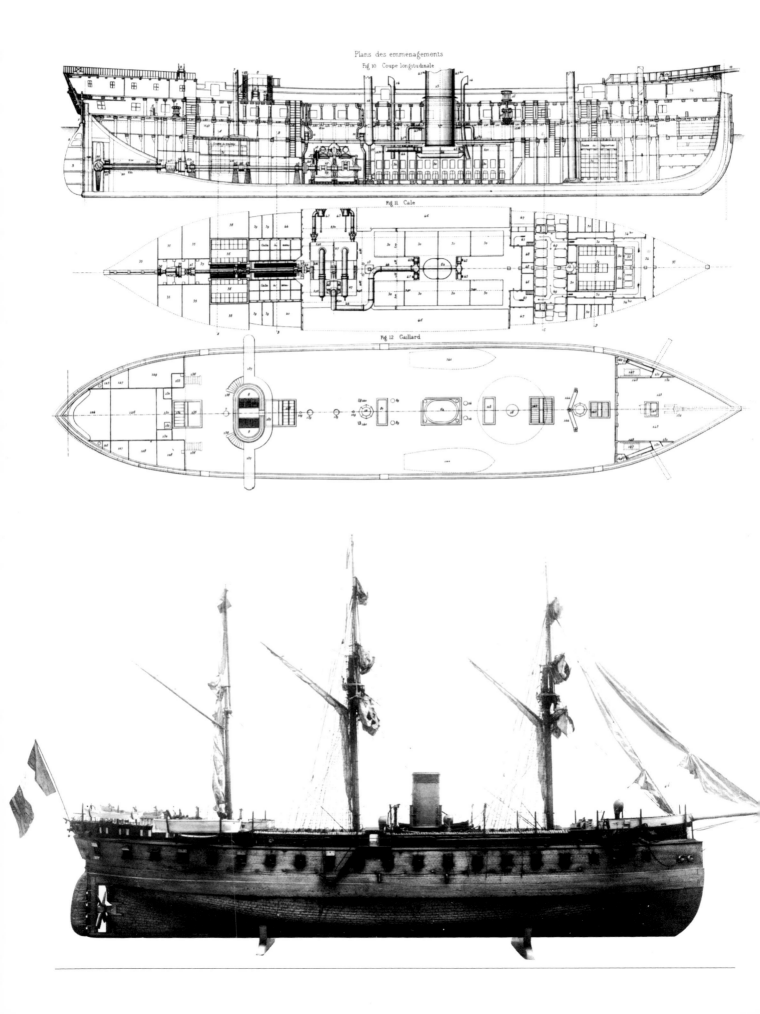

Plans des emmenagements
Fig 10 Coupe longitudinale

Fig 11 Cale

Fig 12 Gaillard

only casualties suffered were from splinters entering through the embrasures or the overhead hatches.

The Crimean War came to an end before the floating batteries could prove themselves in a massive assault on Kronstadt planned for the spring of 1856. With it ended very quickly the brief *entente* between Great Britain and France for almost immediately the government of Napoleon III embarked on a program to expand the navy. By March 1858 preparations had been completed for building four ironclad battleships and the press was full of heady speculation about the end of British naval supremacy. Needless to say the British did not take kindly to talk of Imperial Cuirassiers stabling their horses in Westminster Abbey, and with the lessons of Sinope, Sevastopol and Kinbourn very much in mind, looked for some answer to the threat.

The French were lucky in having the leading naval architect of the day, Dupuy de Lôme. He took the design of his outstanding steam two-decker *Napoleon*, built in 1850, and modified it to carry iron side-plating capable of keeping out the newest 16-centimeter (6.5-inch) explosive shell fired by the 55-pounder muzzle-loading rifled gun. What resulted was hardly a beautiful ship, for *La Gloire* was squat and ugly, but she was soundly executed and fully justified the fame of being the world's first armored *seagoing* battleship. She was to be followed by two sisters *Invincible* and *La Normandie* and a slightly larger ship, *La Couronne*.

For a short while the British appeared to be paralyzed by the specter of resurgent French naval power, and a parliamentary committee was able to recommend nothing more than further conversions of three-deckers to steam power. But behind the scenes there was feverish activity. During the summer of 1858 tests were carried out on samples of iron plate fitted to the side of the old ship of the line *Alfred*, and against the iron-hulled floating battery *Erebus* and her wooden-hulled sister *Meteor*. No fewer than a dozen shipyards and designers were asked to submit designs for ironclads, for the one trump card the Admiralty could play to beat the French was the ability of the British shipbuilding industry to outbuild the French. In November 1858 the Naval Estimates presented to Parliament included a sum for the construction of two armored 'frigates.' The choice of the word frigate was not a euphemism. In 1854 the United States Navy had laid down five big frigates, and two years later the RN had laid down six in reply; these single-decked ships had impressive speed and gun power (but no armor) and would have

been a serious threat to a sailing three-decker or even a steam-powered one. For reasons of topweight it would not have been possible to give *La Gloire* two gundecks, but already opinion had swung round in favor of big frigates, and the idea of an armored frigate being the new capital ship was not quite as revolutionary as it might have been a few years earlier.

Once mobilized the mighty British shipbuilding industry swung into action with its usual efficiency. The first ironclad, the *Warrior* was laid down at Blackwall on the River Thames in May 1859 and launched on 29 December 1859, not long after the completion of *La Gloire*'s trials. The *Black Prince* was slightly later, being launched on the Clyde in February 1861, and the two ships entered service in October 1861 and September 1862 respectively. They were not only seagoing ironclads but also the world's first iron-hulled ironclads, for the British had already built a class of iron-hulled floating batteries in 1855–56 and in addition had the best-equipped iron shipbuilding resources. In any race of this sort the French could not compete for they lacked

Above: The graceful frigate stern and single propeller of HMS *Warrior*. Her fine lines gave her more speed than *La Gloire*.

Top left: Inboard profile and deck plans of *La Gloire*, the first seagoing armored ship.

Bottom left: *La Gloire* had a short, tubby hull but she was a competent design based on the successful *Napoleon*.

the industrial capacity. Their first iron-hulled ship, *La Couronne* was laid down more than a year ahead of the *Warrior* but was launched a month after her.

There were significant advantages in using iron. Most important was the possibility of providing watertight bulkheads, but it also produced a much stiffer hull which could carry the weight of heavier guns as well as the armor.

The hull could be made longer for high speed without undue risk of hogging and sagging as wooden hulls were likely to. Another weakness of a wooden hull was the need to keep the gun-ports close together (because the hull had to be kept short), which made it easier for a hit from a large-caliber shot to punch in the small armor plates and so disable several guns in the battery with a single shell.

Comparison of La Gloire and HMS Warrior

	La Gloire	Warrior
Displacement:	5617 tons	9000 tons
Length:	252 feet	367 feet 6 inches
Machinery:	1-shaft reciprocating engines, 900 ihp	1-shaft trunk engines, 1200 ihp
Speed:	12 knots	14 knots
Armament:	26 55-pdrs (16-cm)	26 68-pdrs (8-inch) 10 110-pdrs (7-inch) 4 70-pdrs (4.7-inch)

The appearance of the slim and graceful *Warrior* and *Black Prince* in the Channel Squadron in 1862 did much to dispel the panic and helped to restore Anglo-French cordiality to some extent. The 'black snakes of the Channel' created a tremendous impression for unlike *La Gloire* they were remarkably handsome ships. Nor were they the only British ironclads; by 1866 another nine were completed, during which time the French navy had completed only two. Not only did French arsenals and shipyards lack the capacity for building rapidly but the demands for increased expenditure on the army tended to siphon off the money which could have been spent on modern equipment for rolling armor plates

and building steam machinery.

Neither *La Gloire* nor *Warrior* enjoyed their prestige for long. The following classes were even bigger and better protected, so big in fact that they were too unwieldy. Clearly the broadside ironclad could not develop much further and so designers turned to other means of providing the gunpower needed. One such idea was the cupola or turret, proposed in 1854 by John Ericsson, a Swedish engineer and also stumbled on in the Black Sea in 1855 by a Royal Navy captain, Cowper Coles. Coles had put a 32-pounder gun on a crude turntable on a raft, nicknamed the *Lady Nancy*, and in 1859 he produced plans for a gun raft protected by an iron shield or cupola. The navy liked his

Below: Sweden honored John Ericsson by naming a monitor after him in 1865. She and her sisters *Thordon* and *Tirfing* were very similar to the American types.

ideas and installed a prototype turret on the deck of the floating battery *Trusty* for firing trials in 1861. It was hit 33 times by heavy shells but continued to function. The Admiralty was convinced and drew up plans for turret-armed ironclads.

Events overtook the Admiralty's plans for in 1861 the American Civil War broke out. The new Federal Navy's problems were acute, for not only had it lost the cream of its officers to the South but also by April 1861 its main dockyard at Norfolk, Virginia. The Confederates found the new steam frigate *Merrimack* (one of the five ordered in 1854) lying burned out and scuttled in the dockyard, but when the waterlogged hulk was raised they were delighted to find that the machinery was still in good condition. The problem faced by the Confederate Navy was the reverse of the British and French position in the Crimean War; it was blockaded in its harbors by Union ships and under bombardment from land artillery. If the *Merrimack* could be turned into an armored battery she would be able to run the gauntlet of the Union artillery and smash her way through the unarmored blockading squadron.

By June 1861 work had started at Norfolk on rebuilding the ship, using railroad iron to build up a sloping casemate housing a battery of 10 guns (two 7-inch, two 6.4-inch and six 9-inch). Although renamed CSS *Virginia* she remained better known by her original name, and the work on her was completed by the early spring of 1862. The Federal authorities, aware of her existence, had authorized the construction of ironclads to match her in August 1861 but the problem remained of what to build. Fortunately the Swedish engineer John Ericsson was on hand with his ideas for a turret-ship (last heard of in 1854). For want of anything better the Navy Department accepted his proposals promptly, and as speed was essential Ericsson simplified the design as much as he could. His

solution was the antithesis of the battery ship, an armored raft surmounted by a single revolving turret containing two 11-inch guns. Her name was to be *Monitor*, 'that she might serve as a warning to others.'

It was a race between the *Merrimack* and the *Monitor*, but the Federal shipyards were up to the task and the *Monitor* commissioned on 25 February 1862, just four months after the contract had been signed. Even so she was nearly too late, for on the day she was due off Hampton Roads the *Merrimack* put to sea. The Confederate ironclad destroyed the wooden frigates *Cumberland* and *Congress* with apparent ease, showing just how helpless an unarmored sailing warship was against an ironclad. The *Cumberland* suffered 121 dead out of a total complement of 376 and the *Congress* 240 killed out of 434.

Next morning the *Merrimack* reappeared, intent on finishing off the big screw frigate *Minnesota*, which had escaped destruction the previous day by running herself aground. At

Left: John Ericsson, engineer, ship designer and inventor of the USS *Monitor*, which gave her name to a revolutionary type of ironclad.

Below: Midship section of the USS *Monitor* through the twin turret, showing the raft construction.

Bottom: Officers of the *Monitor* relax on deck in the James River in July 1862. The dent in the turret was made by a shell from the *Merrimack*.

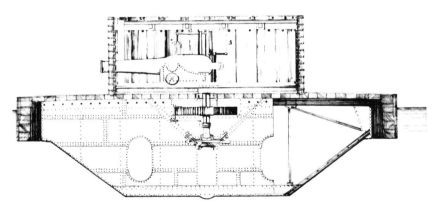

Above: The *Catskill*'s emergency steering position in the engine room.

Top: An officer's cabin aboard the monitor *Catskill*, with sunlight shining through an open skylight.

first she ignored the *Monitor*, mistaking her for a water-tank when she was sighted alongside the *Minnesota*, but as soon as the 'cheesebox on a raft' opened fire with her 11-inch guns the *Merrimack* switched targets and tried to dispose of this impudent intruder. Firing continued for about three and a half hours, the *Monitor* firing her guns every seven or eight minutes and the *Merrimack* taking 15 minutes for each broadside. The baffled Confederates eventually tried to ram but the little turret ship could turn in one sixth of the length needed by the clumsy *Merrimack*. When several attempts failed the *Merrimack* withdrew to her anchorage in the James River, and the Battle of Hampton Roads was over.

The two doughty opponents met again on 11 April but did not engage as Admiral Goldsborough had strict orders not to risk the *Monitor*. Just what might have happened if the

Monitor had used a heavier charge than the very light 15 pounds of powder allowed for her unproofed guns, or if the *Merrimack* had fired solid shot against *Monitor*'s armor can only be guessed at, but the result of the action on 25 February confirmed the value of both armor plating and the revolving turret. Although often claimed to have been the reason why other navies became enthusiastic, Hampton Roads merely provided public proof. Some days before the battle the British Admiralty had in fact given approval to build two turret ships; Admiralty records mention their Lordships' pleasure at the news of Hampton Roads, for it meant that their decision was less likely to be publicly disputed. One hull was to be a four-turret coast-defense ship built of iron but the other was to be converted from an incomplete 131-gun 1st Rate to test the feasibility of converting a number of the big wooden hulls in various stages of building. The *Prince Albert* and *Royal Sovereign* were neither handsome nor famous but they embodied certain characteristics which were years ahead of their time; multiple center-line turrets and no reliance on sails other than for steadying the ship and 'easing' the engines. Both ships were successful in their modest role but were soon overshadowed by even more radical designs.

On the other side of the Atlantic there was an understandable craze to build more and more monitors, for the little ship gave her name to a new type. The Southern States could not match this with their primitive industries, but succeeded in building a second *Virginia* to replace the *Merrimack* (burned in May 1862) and the *Tennessee*. The *Virginia* was scuttled when Richmond was abandoned but the *Tennessee* fought Farragut's fleet at Mobile Bay in 1864, surrendering after being disabled by the monitor *Manhattan*. The monitors proved very useful on the big rivers but they were barely seaworthy; the *Monitor* herself had very nearly foundered on her way down to Hampton Roads and did actually founder in a gale off Cape Hatteras at the end of 1862, and the *Weehawken* sank in 1865. Although much bigger monitors were built, culminating in the 4440-ton *Dictator* armed with two 15-inch guns, they were no substitute for a seagoing fleet. Anti-British feeling in the North ran high as long as sympathy for the Southern cause persisted in Britain, but wild talk of sending the monitor fleet to settle scores with John Bull was ludicrous for they could never have fought their guns in the Atlantic.

Although most navies built monitors of various kinds the British were already moving to the next step, an ocean-going turret ship. In 1866 they ordered the *Monarch*, an 8000-ton

fully-rigged ship armed with four 11-inch guns in twin Coles turrets. Ever since the Battle of Hampton Roads the inventor had demanded that a turret battleship should be built, but when he saw the *Monarch* he still was not satisfied. Being a gunnery officer he could not accept her high freeboard, and wanted the lowest possible freeboard to give his guns a 360 degree arc of fire. The Admiralty supported the Chief Surveyor, Edward Reed, in his claim that it was impossible to reconcile Coles' ideas with adequate seakeeping and stability, but they reckoned without public opinion. By carefully orchestrating parliamentary and press support Coles was able to persuade the First Lord to let him design a second turret ship, to be called HMS *Captain*.

Not unnaturally Edward Reed, widely regarded as one of the two greatest living naval architects, took this as a slur on his department's competence, especially as Coles had no

technical qualifications. The official attitude was that Coles would have a free hand to advise and consult with the builders, the Laird Brothers, but the customary supervision of construction by Admiralty Overseers would not be allowed. In such an atmosphere of discord the *Captain* took shape, and when she was launched in March 1869 it was discovered that she was considerably overweight. Coles was not particularly perturbed and showed no inclination to reduce the great spread of canvas the ship was to carry. Finally she went to sea in January 1870, and to show official approval the First Lord of the Admiralty, Mr Hugh Childers, announced that his only son would sail in her as a midshipman.

All went well for the first three months and the ship's officers expressed great satisfaction with her. However, on the night of 6 September 1870, while the Channel Squadron was beset by a gale, HMS *Captain* was swept over on her beam ends and then capsized. So rapid was the disaster that only seven out of the 473 men aboard escaped. Among those who went down with her were Captain Coles and the midshipman son of Hugh Childers.

In the uproar that followed a searching enquiry revealed that HMS *Captain* had foundered from the 'pressure of sail assisted by the heave of the sea.' Modern naval architects would agree with that but go further in blaming the unusual form of hurricane deck, which acted as a wind-trap when the ship rolled excessively. Edward Reed was amply vindicated when his *Monarch* proved an outstanding success, but what was really at stake was the need to take professional ship-designers seriously. Although there have been subsequent attempts by naval officers to meddle in ship-design there have been no more *Captain* disasters because the details of stability have been entrusted to trained people. The process was already moving slowly and the loss of the *Captain* did no more than hasten it, but 1870 marks the end of the first experimental phase in the evolution of the battleship.

Above: The only US monitor with three turrets, the USS *Roanoke*, had been a frigate of the same class as the *Merrimack*. She was cut down and rebuilt in 1863.

Left: The French *L'Océan* (1870) carried four 10.8-inch guns in her central battery and steamed at 13 knots. Her strengthened ram bow was considered part of her armament.

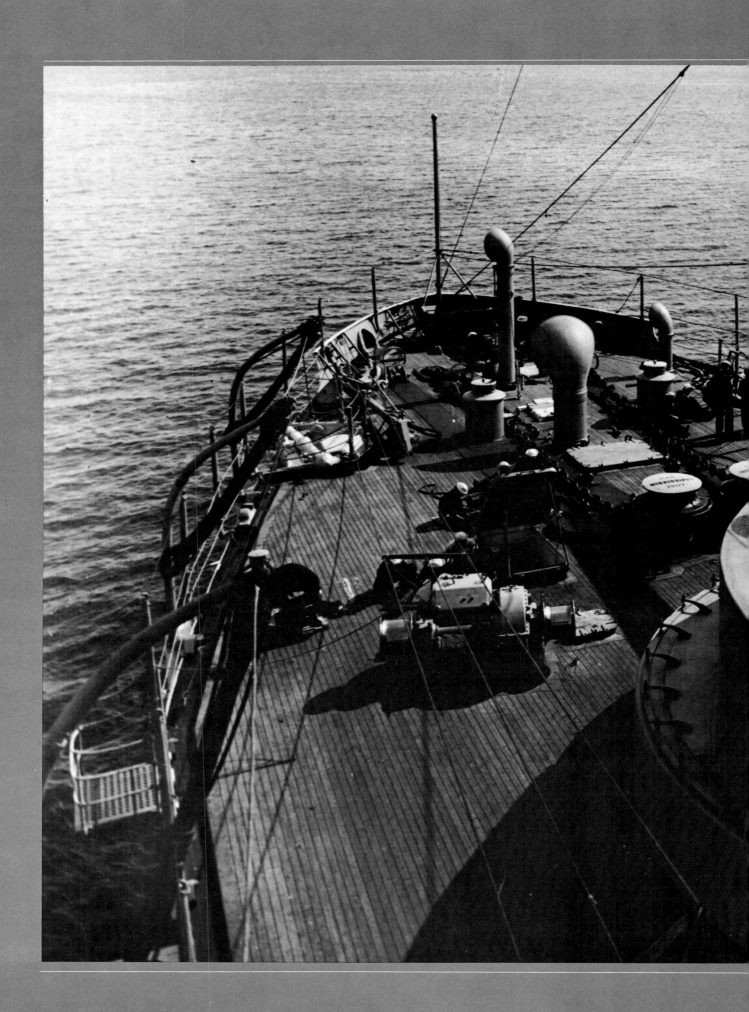

2. GUNS AND ARMOR

The *Captain* disaster did nothing to slow up developments, but throughout the 1870s navies built a bewildering variety of types in an effort to find the best solution. The spur was the need to protect ships against the newest guns, for with metallurgy making rapid advances it was just as easy to improve guns as it was to improve armor. As early as 1853 the British had introduced a rifled gun, the oval-bore Lancaster 68-pounder, and two years after that the French *système la hitte* achieved much the same thing by fitting shells with studs which fitted into shallow grooves. These were both muzzle-loading guns, in which the shell was forced into the rifling by the explosion of the charge. Their worst drawback was the amount of windage, the gap between the shell and barrel, needed to allow the shell to be loaded.

Breech-loading (standard until the end of the sixteenth century) offered the chance to overcome windage and in 1859 William Armstrong demonstrated his system to the Admiralty. He used a removable breech-block to close the chamber behind the charge and then rammed it tight by means of a hollow screw (through which the shell and charge were passed when loaded). There was strong support from press and Parliament once again (Armstrong was even better at publicity than Cowper Coles) and the Armstrong gun was adopted promptly the same year. But once again things went wrong, not as disastrously, but demonstrating just as clearly what happens when public pressure takes over from professional judgment.

The disaster occurred during the bombardment of Kagoshima on 15 August 1863 when a British squadron tried to exact an indemnity from the Daimyo of Satsuma for the murder of an English merchant. The action was just another of the series of minor bombardments of the gunboat diplomacy era but it was distinguished by a number of gun accidents which showed that the Armstrong was unreliable. In an action lasting two hours 21 Armstrong guns fired 365 rounds and suffered an aggregate of 28 accidents, the worst being when a 7-inch gun in the flagship blew out its breech-block and concussed the entire gun crew.

The answer was to change to the French system of 'shunt rifling' and muzzle-loading, for the Royal Navy could not afford any more expensive experiments which might involve reequipping the entire fleet. The French themselves had decided to change to breech-loading but they had adopted a hinged threaded block, without the separate breech-block which had caused all the problems in the Armstrong gun.

The threaded block took so many turns to open and shut that the logical improvement was to cut away every sixth part of the threads; it needed only a one-sixth turn to shut it and yet it retained the strength of the fully threaded block. Friedrich Krupp preferred to use a sliding block but all three groups of manufacturers, British, French and German, now strengthened their gun barrels by shrinking hoops on them, as first proposed by Armstrong.

Guns now got much bigger. At the end of the Crimean War the biggest gun afloat was a 10-inch 84-pounder. Armstrong produced a 13-inch 600-pounder and at the end of the American Civil War the monitor *Puritan* was going to receive two 20-inch Dahlgren guns. The breech-loaders also grew rapidly in size and weight; the 7-inch 110-pounder Armstrong weighed a mere 4 tons whereas the 13-inch 600-pounder weighed 22 tons, causing many more problems in handling. A whole variety of devices came into use to control the recoil of such monsters and steam and hydraulic training were developed for the mountings.

The French did not favor the turret, preferring to develop the barbette, a circular iron shield inside which the guns revolved. Its advantage was that it weighed less than the turret since the training machinery was only turning the guns rather than the whole mass of the armored turret. In turn this meant that the guns could be carried higher above the waterline than in turret ships, and from the end of the 1860s a number of French ships were fitted with four single guns in barbettes disposed lozenge fashion (one forward, one aft and one port and starboard amidships). The British found a different solution in the central battery, shortening the side armor in order to give heavier protection to a central casemate amidships. This arrangement not only saved length but with recessed gunports allowed the guns to fire when pointed closer to the center line. One of the weaknesses of the older broadside ironclad was this lack of end-on fire, and many expedients were adopted to provide chase guns at both bow and stern.

The biggest problem was lack of experience on which practical design ideas could be based. The Civil War was recognized to have been a special case never likely to recur, while the Crimean War had even less relevance. But on 20 July 1866 the Austrians and Italians fought off the Island of Lissa in the Adriatic, and this action was to have as much immediate influence as Hampton Roads. Not only was it the first full fleet action fought in European waters since Trafalgar but also the first battle involving seagoing ironclads. The newly unified

Previous page: The forward twin 12-inch turret of the USS *Mississippi* in 1908. Until the twentieth century smartness of drill and appearance were the principal criteria of a battleship's efficiency.

Italian state had joined Prussia in furthering Bismarck's plans to acquire part of the Austro-Hungarian Empire. The Austro-Hungarian fleet under Rear Admiral Wilhelm von Tegetthoff had only seven armored ships out of a total of 27, against Count Carlo Pellion de Persano's 12 out of a total of 34. The Italians had spared no expense to create a new fleet out of nothing, including a new turret ram, the *Affondatore*, and could bring 200 modern rifled guns into action against the 74 in Tegetthoff's fleet.

Faced with such opposition Tegetthoff could either avoid action or trust in superior seamanship and tactics. He chose the latter, knowing full well that Persano was an indecisive leader with poorly trained men. He ordered his ships to close the range so that the older muzzle-loaders could penetrate the Italians' armor and instructed them to ram enemy ships whenever possible, to throw the Italian battle line into confusion: 'ironclads will dash at the enemy and sink him.' Tegetthoff's three divisions attacked in an arrowhead formation and he achieved his primary objective by slipping with his armored division through a gap in the Italian line. Ramming proved almost impossible because of the clumsiness of the ships and the dense clouds of powder-smoke but the attempts produced a fierce melee. The *Affondatore* closed with the steam two-decker *Kaiser* and riddled her upperworks with 300-pounder (10-inch) shells from her twin turret but failed to ram her. The wooden 90-gunner suffered one hit which wounded six men and dismounted a gun but she immediately tried to ram the big ironclad frigate *Re di Portogallo*. The *Kaiser* lost her bowsprit, foremast and funnel, and was then hit by a single 10-inch shell and 13 more from 64-pounders (6.3-inch) in quick succession. With 61 casualties and a fire blazing forward she clawed her way clear to have time to remove the tangle of wreckage on the upper deck and fight the fire.

The decisive moment came soon after, when Tegetthoff in his flagship the 18-gun ironclad frigate *Ferdinand Max* sighted the *Re d'Italia* through a gap in the smoke. The big Italian frigate had been disabled by a shell-hit in the rudder and could do nothing to escape her fate as the *Ferdinand Max* bore down on her. The iron spur under the forefoot struck the *Re d'Italia* full amidships and tore an enormous hole on the waterline; then the *Ferdinand Max* reversed her screws and pulled away, making the hole even larger. The doomed Italian ship sagged slowly over to starboard as hundreds of tons of water rushed into her hull, righted momentarily and then rolled the other way and capsized, taking 622 men with her. The only other casualty of the battle was the small ironclad *Palestro*, which blew up after being set on fire by the *Ferdinand Max* early on, but the Italians had had enough. After a show of bravado by hoisting a signal for 'General Chase' Persano withdrew to Ancona and left the field to Tegetthoff.

The most extraordinary thing about Lissa is that casualties were comparatively light: 682

Below: US warships of the 1890s at sea. Left to right, the gunboat *Marietta*, the monitor *Puritan*, the battleships *Illinois* and *Iowa* and the torpedo boat *Stringham*.

killed and 153 wounded on the Italian side and 28 killed and 138 wounded on the Austro-Hungarian side. Tegetthoff's ships had fired 2786 shot and shell and yet caused only 44 casualties, and the Italians needed 1400 shells to inflict 166 casualties, 61 of them in the wooden *Kaiser*. The real lesson was that poor maneuverability and inaccurate guns made it very difficult to sink ships in battle but the world's navies seized on the ramming of the *Re d'Italia* as proof that the ram was a potentially decisive weapon. For another 30 years battleships would be built with massive reinforced stems and practice ramming tactics, achieving little apart from spectacular collisions with their squadron mates. And yet we can now see that Lissa actually proved the opposite. Over and over again ramming attempts by Tegetthoff's ships had been defeated, and the solitary success had only been achieved because the victim was unable to steer. Even more astonishing was the performance of the *Kaiser* after her first engagement with the *Affondatore*. She re-engaged the Italian turret ship, trying to ram her three times, and when set on fire once more withdrew to San Giorgio to effect temporary repairs. She had fired 850 rounds and had been hit 80 times, suffering 24 killed, 37 severely wounded and 38 slightly wounded. Clearly the effects of shellfire on wooden ships had been greatly exaggerated, and provided they were manned by disciplined crews they could survive.

With gun design and metallurgy still in a state of flux there was very little to help the designer in formulating requirements and a confused and contradictory battle like Lissa was very little help. The two lessons learned, the need for end-on fire and a strengthened ram bow, were the wrong ones, but indirectly they proved a positive influence. More emphasis was placed on handiness and inevitably the revolving turret was vindicated as it offered the widest arcs of fire. But there were to be many quaint compromise solutions to the problem during the 1870s: turret rams, casemate ships, box-battery ships and more of the monitor type. All of them looked like the freaks that they were, often with huge exaggerated rams. There were ships with two turrets side by side on the forecastle, turrets *en echelon* amidships and combinations of central batteries and barbettes, in a bewildering profusion of types. There was no major naval war nor even a serious threat of a war (in Europe at any rate) and this 'fleet of samples' was as much a reflection of the industrial growth of Europe as a response to any specific threat.

Battleship design, with its need for ever harder armor plate, bigger forgings for guns and the most powerful machinery, made heavy demands on technology. At first the British had dominated the field, followed by the French, but both Germany and Russia soon developed the capacity to build ironclads. For the time being the other industrial giant, the United States, was content to ignore these developments, for the aftermath of the Civil War focussed energies on developing internal resources rather than maritime power. There was also the understandable tendency to assume that the large fleet of monitors was sufficient investment in sea power for the time being. The magnitude of that error would become apparent later, but without the threat of a naval war to disturb the complacency of the politicians there was little that anyone could have done to change matters.

It was left to the British to show the way ahead. In 1869, while the *Captain* was still fitting out and the *Monarch* had just gone to sea, Parliament voted for three more large turret ships. With the reputation of the US

Navy's monitors standing very high, Sir Edward Reed chose a low freeboard ship with two large twin turrets protected by an armored 'breastwork' and no sails of any sort. The arrival of the USS *Miantonomoh* at Portsmouth in 1867 lent some credence to American claims of the big monitors' seaworthiness but eyewitness accounts testify to the hellish conditions aboard when she first arrived. Then came the disaster of the *Captain*, a year after the keels of the first two of the new turret ships, and a storm of protest arose. To settle doubts about their design the First Lord of the Admiralty appointed a Committee on Designs. When it met in January 1871 Reed had already resigned as Chief Constructor, to take up a job in industry, and so the committee had little difficulty in recommending some changes to the design of the new ships. But what was much more important was the recommendation that sail power should be abolished for large warships. It was recognized that a 'very high degree of offensive and defensive power' could not be combined with real efficiency under sail. Although later ships were to carry masts and yards they were there to save coal on long cruises by easing the engines; the age of fighting under sail was over.

HMS *Devastation* went to sea in the spring of 1873 under a cloud of suspicion and pessimism never seen before or since. An anonymous hand even placed a notice by the gangway on the day of her commissioning: 'Letters for the *Captain* may be posted here.' But she confounded the critics when she was sent out into the eastern Atlantic with the old broadside ironclad *Agincourt* and the new central battery ship *Sultan*. She behaved as well or better in a variety of sea states and showed that she was in no danger of capsizing. Naturally her low freeboard imposed some limitations, for it was impossible to prevent water from finding its way down below the forecastle, but there was no doubt that she could steam and fight in the Atlantic. She and her sister *Thunderer* also carried nearly 2000 tons of coal and could steam 5000 miles at cruising speed. The lack of sail power gave the

Below: An imaginative rendering of the Battle of Hampton Roads, 9 March 1862. Based on a French officer's sketch, the most serious inaccuracy is showing the frigate *Congress* still afloat.

Above: The broadside iron-clad HMS *Agincourt* was launched in 1865 as a five-masted 50-gun frigate but later had two masts removed.

Top right: The bare, compact superstructure of the *Devastation* is in marked contrast to the fully rigged ironclads.

Center right: Central battery ships like HMS *Sultan* (1871) were an improvement on the broad-side ironclads, with a shorter armor belt and limited ahead and astern fire.

Bottom right: The *Inflexible* (1881) tried to combine a short length of armor with ahead and astern fire by putting two turrets *en echelon* amidships, but the age of the freak battleship was nearing an end.

guns maximum arcs of fire ahead and on the beam, and reduced complement considerably. Without the weight of heavy masts and standing rigging the hull was also steadier and a greater proportion of weight could be devoted to armor, coal and habitability. In the words of a contemporary commentator the design was an 'impregnable piece of Vauban fortification with bastions mounted upon a fighting coal mine.'

Developments in guns were to reduce the *Devastation*'s 10-inch to 12-inch iron armor to no more than piecrust within a few years but she stands out as the most effective concept of the 1870s, a balance of all the fighting qualities. It was to be another 10 years before a similar balance was achieved anywhere in the world, for the naval world soon plunged into yet more confusion. The biggest factor in this confusion was a series of 'monster guns' produced by the British company Armstrongs. In 1872 the Italian navy laid down two large battleships, the *Duilio* and *Dandolo*, intended to have the thickest armor and the largest guns. The designer, Benedetto Brin had considerable talent and his first proposal for ships with four 38-ton 12.5-inch guns would have given the Italians two ships well able to deal with any opposition in the Mediterranean.

The problems started when Armstrongs offered the Italian navy their latest 60-ton 15-inch. No sooner was the offer accepted than Armstrongs raised the bidding with a 100-ton 17.7-inch, and Brin was told to alter his design

accordingly. He was forced to recast the design entirely, choosing a raft body underwater, on which the ship would be able to float if the remaining two-thirds of the ship was flooded. The basic concept was sound enough but the immense size of the guns made nonsense of it. The light hull could not take the stresses of continuous firing, and as the 100-tonners fired so slowly and inaccurately the ship could easily be riddled by shells from lighter and faster-firing guns before she could reply. The Italian answer to these criticisms was beguil-ingly simple; they knew that the *Duilio* and *Dandolo* would not show up too well in a ship-to-ship combat but if given sufficient speed they might be able to stay out of trouble.

The 'monster gun' produced a panic in Britain very hard to comprehend today, and in 1874 the official reply, HMS *Inflexible*, was laid down. She resembled the *Duilio* in having the turrets *en echelon* amidships, a raft body and central citadel protected by 24 inches of compound, steel and iron armor. The Ad-miralty did not try to match the Armstrong 100-ton gun, preferring the Woolwich 80-ton 16-inch. This gun marked the peak of rifled muzzle-loader development for the barrels were too long to be run back into the turret for loading. Instead they depressed the muzzles below an armored glacis on the deck to allow first the charge and then the shell to be rammed upward into the barrels.

Inflexible was a triumph of mechanical ingenuity, the first battleship lit by electricity,

the first with anti-rolling tanks, the thickest armor ever, and above all, the first design to be thoroughly tested and discussed before building started. And yet she was a failure: because there was no Naval Staff nobody had thought out any task for her to perform. Her sole claim to fame was to be commanded by Captain John Fisher at the bombardment of Alexandria in 1882. She fired 88 rounds and was hit by a single 10-inch shell which inflicted considerably less damage than the blast of her own 16-inch guns.

Undaunted by the shortcomings of the *Duilio* and *Dandolo* Benedetto Brin went on to build a much improved version, the *Italia* and *Lepanto*. This time the guns were 103-ton 17-inch breech-loaders, and instead of two turrets he sited them in a heavily armored redoubt resembling the French barbette system. They were also technical masterpieces but were overtaken by two events, the introduction of high explosive shells and quick-firing guns. Now it was possible to inflict great damage on unarmored structures with comparatively light guns (mostly 4-inch to 6-inch). Ships like the *Italia* and *Inflexible* would be put out of action even if they could not be sunk.

There was a revolution brewing in big guns as well. In 1878 Sir Andrew Noble and Professor Abel began a series of experiments which proved that ballistics would be improved with slower-burning gunpowder. This in turn called for longer barrels to ensure that the powder burned for as long as possible. Other improvements were to make the gunpowder in large or 'pebble' grains and to chamber the breech to improve combustion. All this led to agitation for a return to breech-loading, for longer guns were very hard to load from the muzzle-end. The *Inflexible* arrangement showed that the limit had been reached, and chambering of muzzle-loaders was difficult. In 1875 Krupp had introduced his mantlering system, shrinking a jacket over the breech-end to provide much greater strength than before, and no muzzle-loader could match the extra power. Excessive parsimony had pre-

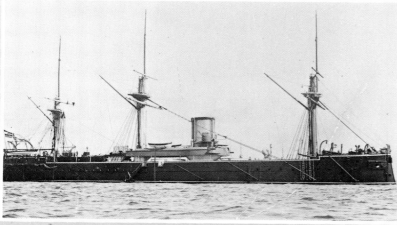

being double-loaded. As described by Admiral Seymour, both guns in the turret were fired together but amid the tremendous noise and concussion the gun-crews did not notice that one of the guns had not recoiled (ear plugs were not known and men often shut their eyes or put their fingers in their ears at the moment of firing). The run-in levers were immediately worked and when the rammer's mechanical indicator showed that the fresh charge had not gone all the way up the barrel it was assumed to be a jam. However, there were now two powder charges and two shells nose-to-tail in the barrel, and when the firing lanyard was pulled an explosion burst the gun, wrecked the turret and killed several of the gun's crew. The tragedy did much to undermine the reputation for simplicity and reliability built up around the big rifled muzzle-loaders and paved the way for the adoption of breech-loading.

In April 1879 a committee was set up to consider a revival of the breech-loader and at

Above: The German turret ship *Friedrich der Grosse* (1877) was armed with four 10.2-inch breech-loaders and steamed at 14 knots.

Top: HMS *Thunderer* in April 1891 after being rearmed with breechloading 10-inch guns and small quick-firers.

Right: The armored corvette *Baden* and her three sisters were built in 1875–83. The six 10.2-inch guns were mounted side by side in a barbette forward and in a square redoubt aft.

vented the Royal Navy from considering such innovations for the best part of a decade, but a growing awareness that the European navies could shorten the British lead loosened the purse strings at last.

Fortunately public opinion went along with the technical need for change. On 2 January 1879 HMS *Thunderer* suffered a bad accident when one of her 12.5-inch guns burst through

the same time a group of technicians went to Meppen to witness trials of the new Krupp guns. They returned full of enthusiasm and the Admiralty wasted no further time in ordering a new 25-ton breech-loading 12-inch. The French interrupted-screw method was adopted rather than the Krupp sliding breech, with several safety interlocks to prevent accidents.

Once the decision was made to revert to breech-loaders the Royal Navy planned a new series of ships to match the latest French barbette ships. The first was the *Collingwood*, designed by a constructor who was to prove the most able naval architect ever produced by Britain and easily the most talented designer of his generation. William White had to cope with an unspoken but nonetheless real political reluctance to permit the displacement of ships to rise above 10,000 tons, and so he had the choice of limiting protection or freeboard. He chose the latter, copying the layout of the *Devastation* but putting the four 12-inch breech-loaders into twin barbettes on the French style. The protection was limited to a

short but thick belt on the waterline, closed at either end by a transverse bulkhead. The ends were left 'soft' or unarmored, on the assumption that as long as the central citadel remained intact the ship would float with both ends flooded. The *Collingwood* turned out to be remarkably handsome, with two tall funnels

Above: The turret ship *Andrea Doria* and her sisters were repeats of the successful *Duilio* and *Dandolo*.

Top: As well as her four 17-inch guns the *Italia* carried eight 6-inch and four 5-inch.

Top: The Russian
Imperatritsa Ekaterina II
(1889) had three twin 12-
inch barbettes, an unusually
heavy armament then.

Above: The Austro–
Hungarian battleship
Babenberg shortly before her
launch at Trieste in
October 1902.

and a single light mast. Although her 12-inch guns looked somewhat less imposing than the older muzzle-loaders they were much more efficient and enhanced her neat, functional appearance. Five more improved *Collingwood*s were built, four with four new 30-ton 13.5-inch guns and one, HMS *Benbow*, with two 111-ton 16.25-inch guns. The decision to go for another monster gun was a mistake, for with one round fired every 3–4 minutes the *Benbow* was less likely to score a hit than a ship armed with twice as many lighter guns.

The 1880s and 1890s saw a remarkable upsurge of public interest in naval affairs in Great Britain, the United States and throughout Europe, with naval yearbooks and journals providing more information about warship designs, numbers of guns and fleet strengths. The debates about design had hitherto been confined to administrators and technicians, except when a disaster brought the problem into the public arena, but from now until World War I respective naval strengths became a matter of public concern. Newspapers and books bombarded their readers with information about trials of new guns and armor plate, and as all the underlying tensions came closer to the surface a note of strident

nationalism pervaded it all. The closing years of the nineteenth century became the Age of the Battleship, with rival strengths reckoned in numbers of battleships, but little thought given to broader strategical and technical matters.

There was, for example, the question of the Whitehead torpedo, a self-propelled explosive device which could blow a hole in a ship's side. The Royal Navy had enthusiastically snapped up the invention in the early 1870s without thinking too hard about how best to use it, but

Above: As the century wore on designs became more balanced and symmetrical, like the German coast-defense ships of the *Siegfried* Class, built between 1888 and 1896.

Below: The Chinese turret ship *Chen Yuan* after her capture at Wei-Hai-Wei by Japanese forces during the Sino–Japanese War of 1895.

by the mid-1880s small fast torpedo boats had been built in sufficient numbers to threaten any battleships trying to lie off an enemy harbor. This made any repetition of the classic blockade of the French Brest Fleet by Cornwallis a thing of the past, and some supporters of the torpedo boat even dared to predict that the battleship had no future.

There was to be one more scandal to cast doubt on the design of contemporary ships. On 22 June 1893 two columns of battleships of the British Mediterranean Fleet were maneuvering off Tripoli in the Levant when the flagship, HMS *Victoria* was rammed and sunk by the ship at the head of the second column, HMS *Camperdown*. Controversy has raged ever since over just what went wrong, but the most likely cause was a confusion in Vice-Admiral Tryon's mind about the minimum distance to allow for a half-circle turn. But the alarming fact was that the flagship of the Mediterranean Fleet, holed forward by the *Camperdown*'s ram, turned turtle and sank only 12 minutes later with the loss of half her complement. The cause was quite simple: small openings and leaks made nonsense of the theoretically watertight doors which closed off each major compartment. It was a lesson which would have to be painfully relearned during World War I.

The loss of HMS *Victoria* had no effect on

the development of battleships for her design had already been recognized as unsuitable for further development. Instead William White's *Collingwood* or 'Admiral' Class was chosen as the basis for a new class of big ships to be built under a massive program authorized by the 1889 Naval Defence Act. Growing public agitation about the alleged weaknesses of the Royal Navy had forced the Conservative Government's hand resulting in the appointment of White himself to the job of modernizing the Royal Dockyards. By the end of 1888 White had turned them from inefficient organizations capable only of undertaking repair work into the cheapest and fastest building yards in the world. Had the critics known, the British position was not as bad as they had assumed, for the French Chamber of Deputies had been told in 1886 that only 10 battleships were ready to go to sea. Out of six ships started between 1878 and 1881 only one was nearing completion and another was less than 40 percent complete after five years of work. The French still had most of their timber-hulled ironclads from the 1860s on the strength whereas such ships as HMS *Warrior* had been all but written off.

White's new design, the *Royal Sovereign* Class, matched the scale of the program. With no 10,000-ton limit to circumscribe him he was able to give them the freeboard which the 'Admirals' lacked and to protect half the 6-inch guns against light shells. They were barbette ships, carrying their four 13.5-inch guns in open-topped mountings, but as a sop to the objections of the elderly First Sea Lord, Sir Arthur Hood, an eighth unit was built with turrets instead of barbettes. The case was finally proven; the *Hood* had less freeboard and so could never make the speed of her half-sisters in bad weather.

The impetus given by the Naval Defence Act was not lost, and in the next class of nine ships, the *Majestic*s, White produced his masterpiece. He kept the layout of the *Royal Sovereign*s but provided the barbettes with armored hoods – the term 'hooded barbette' eventually became 'turret' when the old pillbox type had disappeared – and put all the 6-inch guns into armored casemates. They were even better looking than the previous ships, so much so that the US Navy paid White the supreme compliment of copying the look for their *Alabama* Class. It was not only the fighting qualities of the *Majestic* Class that amazed foreign navies; *Majestic* was built by Portsmouth Dockyard in under 22 months, a world record. In less than 10 years the Royal Navy was strengthened by no fewer than 16 first-class battleships, whereas France and Germany completed only 6, and Russia and the United States 4 each.

The French navy went into comparative

Above: The *Re Umberto* and her sisters compared favorably with contemporary designs when conceived in 1883 but were outclassed when they came into service in 1893–95.

Top left: The Greek *Hydra* (1889) was an unusual French design with two single 10.8-inch forward and one aft.

Center left: SMS *Kurfürst Friedrich Wilhelm* (1893) and her sisters saw service in 1914–15 but were then laid up as hulks.

Bottom left: William White's *Royal Sovereign* marked not only a giant step forward in efficiency but a new esthetic harmony in appearance which influenced all later designs.

Left: The USS *Texas* (1895), although obsolescent, was the first of the New Navy armored ships authorized in 1886.

decline at the end of the century, partly because of the patent hopelessness of trying to outbuild the British, partly because of mischievious political meddling but largely because of the influence of the 'Young School' or *Jeune Ecole*, younger naval officers who believed fervently that torpedo boats made the battleship obsolete and that destruction of commerce was a more certain way to beat the British than fighting their fleet. But in their place appeared a much more dangerous rival, the German navy. Starting from a coast-defense force its strength was built up by a series of Navy Acts. These were intended to provide a firm basis for replacing obsolescent tonnage but they pandered to nationalist sentiments in a way that mere totals of money could not. They also alarmed the British, who saw clearly that an enlarged German navy with ships best suited to use in the North Sea could only threaten their own position. With jingoism and tension rampant in all the leading nations of Europe it was not the moment to start a

naval arms race but Admiral Tirpitz did not choose to see this. Protesting all the while that the new battleships were essential to protect German interests, he ordered 20 large battleships between 1890 and 1905. The argument was quite simply that battleships were a measure of national prestige, and as Germany was a strong industrial nation she needed a fleet of appropriate size. A second reason was Germany's desperate need for powerful allies; Tirpitz reckoned to have a big enough fleet to make Germany a desirable ally.

Nothing much had been heard from the US Navy since 1866 for the very simple reason that very little had happened since then. Congress refused to sanction any major warships on the grounds that battleships and big cruisers would lead to international adventures. Far-sighted officers and administrators tried as early as 1874 to get some measure of new building put in hand but were only able to get grudging permission to repair old monitors. To circumvent this they resorted to a trick well

Left: Admiral von Tirpitz, creator of the German High Seas Fleet but ultimately the architect of his country's downfall. His blind adherence to naval expansion created an arms race which led to a European War.

Below: White's masterpiece, the *Majestic* Class, remained the basis of British designs for a decade. HMS *Caesar* (1898) is seen in the standard Victorian livery of black hull, white upperworks and buff funnels.

known in the eighteenth century, the Great Repair, and money was voted to repair five Civil War monitors whose wooden hulls had rotted away. In fact five new ships were built under the names *Puritan*, *Amphitrite*, *Mianto-nomoh*, *Monadnock* and *Terror*, but as material had to be ordered surreptitiously, the work took from 17–22 years to complete. In 1887 at long last five new monitors were authorized, but they were hardly an improvement. In 1895 the *Amphitrite* tried to cruise in the Atlantic and had so many stokers laid out by heat exhaustion that she had to anchor until the temperature in her stokehold dropped.

The first battleships allowed to be built were the small second-class units *Texas* and *Maine*

Top: The monitor USS *Amphitrite* (1895) and her sisters were built to placate political prejudice against seagoing battleships. They were useless as fighting ships but survived as tenders until the early 1920s.

Above: Welcoming the victors in the Spanish–American War, August 1898. Left to right *New York*, *Iowa*, *Indiana*, *Brooklyn*, *Oregon*, *Massachusetts* and *Texas*.

(the latter was first conceived as an armored cruiser) ordered in 1886. Four years later three battleships, *Indiana*, *Massachusetts* and *Oregon* were ordered, but only on condition that they were described as seagoing. With only 12 feet of freeboard and 400 tons of coal they were hardly fit for more than coastal defense and the next ship, the *Iowa* was hardly any better. Not until the three *Alabama* Class were laid down in 1896–97 could the US Navy boast a battleship capable of facing the latest European models.

Russia was in a worse state than France, with a slow building rate and a tendency to pack too many features into a small hull. Recognition of some of the basic weaknesses led the Russians to seek French advice on design, and from them they imbibed many theories on ship construction. The Russian sailors and their officers were brave but lacked good professional training, while the innate corruption of the system made the task of reform hopeless.

To an outside observer the front-line navies at the end of the century appeared marvels of precision. There was as yet no suggestion that color schemes should be chosen for anything but appearance, and the long lines of black hulls, white upperworks and buff funnels impressed civilians. After 30 years of peace, spit and polish had been given too much priority over such matters as gunnery and tactics, but we should not underestimate the seamanship and skills of late-nineteenth-century battleship captains. Gunnery was practiced at no more

than 4000 yards, little more than the extreme range of Nelson's 32-pounders at Trafalgar, but this was because the irregular-burning powders produced a wide scatter of shots at extreme ranges. At a range of 7000 yards or more the only thing visible with the naked eye was a plume of smoke from the target's funnels and the 100-foot-high splash of a 12-inch shell.

Much nonsense has been written about officers throwing shells overboard to avoid the dirt and tedium of gunnery drills, but the facts are that by the end of the century the front-line navies were in a state of reasonable efficiency. But it was all to be made to look obsolescent by a veritable hurricane of change which was to transform them for good.

Above: The USS *Kearsarge* in dry dock in 1899.

Below: For small navies with an indented coastline, the coast-defense battleship made sense. The Norwegian *Harald Haarfagre* (1897) was a diminutive of seagoing battleships, with single 8.2-inch guns and a heavy secondary armament.

3. TSUSHIMA

The problem of all nineteenth-century battle-ship designers was that they were denied all but a few tantalizing fragments of experience. All their theories of gun layout and armor distribution rested on the experience of the Crimean War, Lissa and a half dozen minor actions. Not until 1904 did modern warships meet their equals, when the Japanese clashed with the Russians over which nation was to dominate Manchuria. At the end of the war the Russian Pacific Fleet had been sunk at its moorings by long-range gunfire and the Baltic Fleet had been annihilated at the Battle of Tsushima in May 1905, the first full fleet action since Lissa.

Tsushima was unusual in many ways. Paradoxically it was the last 'straight fight' between battle fleets, before long-range fire-control, submarines and high-speed torpedoes were sufficiently advanced to play a major part in tactics. But in other respects it was full of startling novelties. One protagonist, Imperial Russia, was a world power with big industrial resources, the other had been a backward feudal state within living memory. It saw the first use of radio, marked the first use of gunfire at long range, the first use of offensive mine-laying and the first use of destroyers in surface attacks on a fleet at sea. It was also a test of two rival schools of design, for the Japanese were largely trained and equipped by the British, whereas French influence predominated in the Russian navy.

The Japanese had deliberately courted war with Russia for they were determined to stop Russia from establishing a foothold in Manchuria. The powerful fortified base at Port Arthur was a constant thorn in Japanese flesh and when it was announced that the Port Arthur fleet would be increased from seven to 13 battleships by the end of 1905 the Japanese decided to strike first. They had only six modern battleships and it seemed foolish to wait until the Russian forces were doubled.

On the night of 8 February 1904 10 destroyers attacked Port Arthur and damaged two battleships and a cruiser with torpedoes. This foretaste of Pearl Harbor gave the Japanese Fleet only a temporary advantage but what was more important was the fillip to their morale and confidence. The Russians never gained the initiative thereafter, and when the energetic Admiral Makarov was drowned in his flagship *Petropavlovsk* after she had struck a mine, the fighting spirit of the Port Arthur squadron seemed to wither away. Not even the sinking of two Japanese battleships in a mine-field tempted the Russians; an attempt to break out in August 1904 was the last sign of any activity and four months later 11-inch howitzers belonging to General Nogi's besieging army sank the survivors at their moorings.

Meanwhile the Russian Baltic Fleet had begun its incredible odyssey, steaming around the world to try to save Port Arthur and destroy Japanese ambitions for good. Pausing

Previous page: Last minutes of the *Borodino* at Tsushima, as seen by an imaginative contemporary artist.

Below right: The *Tsarevitch* carried her 6-inch secondary guns in six twin turrets, and the 'tumblehome' was an attempt to offset this extra weight.

Below: The 12,900-ton *Tsarevitch* was designed with considerable French assistance to serve as the prototype for the *Borodino* Class.

only to shell British trawlers off the Dogger Bank in a moment of wild panic, Vice-Admiral Rozhdestvensky's fleet plowed on around the Cape of Good Hope. Despite the dispiriting news of the surrender of Port Arthur the Russians continued their seven-month journey and approached the Japanese islands with the intention of either brushing past Admiral Togo's fleet to reach Vladivostok or bringing it to action and destroying it. With eight battleships, three coast-defense ships, three big armored cruisers, six light cruisers and 10 destroyers Rozhdestvensky seemed to have a

Above: The Japanese *Hatsuse* negotiating the Tyne swing bridge in 1901, with her topmasts lowered to clear the high-level bridge. She was sunk by a Russian mine in May 1904.

Below: The *Retvisan* was built for Russia in the USA but fell into Japanese hands at Port Arthur. After repair she became the *Hizen* and served until 1924.

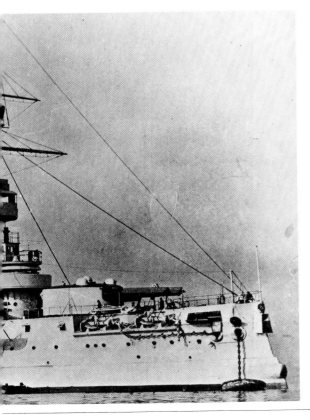

powerful advantage over Togo's four battle-ships, seven armored cruisers and seven light cruisers. But Togo's ships could steam together at 15 knots whereas the Russians had a combined speed of only 12 knots, and there was no comparison between the standards of training in the two fleets.

Early on the morning of 27 May 1905 Togo's scouts sighted the Baltic Fleet entering the Straits of Tsushima but it was in full view before Togo decided to bring his big ships into action. The Russians were already in line ahead, the four modern ships leading, and Togo turned to port to bring his own line

against the head of the line. By 1400 hours the flagship *Kniaz Suvorov* was under heavy fire from several ships and Vice-Admiral Rozhd-estvensky had been wounded badly by a shell splinter. The Russian second division, comprising four older battleships, was under heavy fire from the Japanese armored cruisers and eventually the *Oslyabya* was reduced to a sinking condition by numerous hits.

The flagship drifted away from her consorts, shrouded in smoke from internal fires. A hit from a destroyer's torpedo failed to stop her and three hours later she was still fighting off attacks. At about 1730 hours a destroyer took off the badly wounded admiral but still the smoke-blackened wreck fired her light guns and kept the Japanese destroyers and torpedo boats at bay. Finally at 1920 hours a division of torpedo boats scored two or three hits and ended the agony; the *Kniaz Suvorov* rolled over on her side and went down with all 928

around on a course nearly parallel, making the Nelsonic signal 'The fate of our Empire hangs on this one action. You will all exert yourselves and do your utmost.' The Russians replied with fierce but inaccurate fire at a range of about 6500 yards, which, however, failed to prevent Togo from concentrating his fire

officers and men.

Meanwhile the *Borodino* led the survivors of the battle line in a forlorn attempt to break out of the trap. At about the same time as the sinking of the flagship she was set on fire and her magazines blew up. The *Alexander III* had already capsized after repeated hits from 12-inch shells, leaving only the *Orel* and a handful of older ships circling aimlessly. The Japanese steamed around them firing at will, hampered only by the dense clouds of smoke. When the sun rose next morning the *Orel* and the other survivors could do nothing more and so they surrendered. The only ships to escape were a few which chose internment in neutral ports. It was the first complete victory since Trafalgar, the tactician's dream of a 'battle of annihilation.' Japan was suddenly catapulted to the rank of world power and the Russians reduced (albeit temporarily) to second-class status.

There were many lessons to be learned from Tsushima, but as always as many wrong conclusions as right ones were drawn. The outstanding feature to most commentators was the 'immense' range at which fire had been opened, 7000 yards instead of half that distance, as had been expected. However, all but a few of the more perceptive observers failed to notice that Japanese gunnery had been impressive for its rapidity rather than its accuracy, and only when the range came down

Left: Togo's flagship *Mikasa*. The characters are Togo's famous Tsushima message written in his own hand.

Far left: Admiral Heihachiro Togo, the architect of victory at Tsushima. Although hailed as a new Nelson his somewhat risky tactics were aided by the Russians' tactical ineptitude.

Below: The *Oslyabya* was sunk at Tsushima but her sisters *Peresviet* and *Pobieda* were captured at Port Arthur and became the Japanese *Sagami* and *Suwo*.

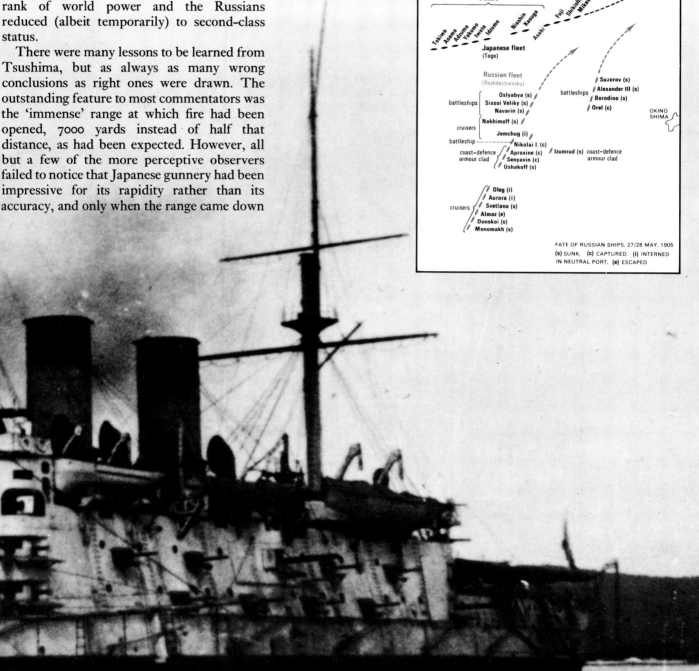

FATE OF RUSSIAN SHIPS, 27/28 MAY, 1905
(s) SUNK. (c) CAPTURED. (i) INTERNED
IN NEUTRAL PORT. (e) ESCAPED

Above: The *Asahi* was the third of four battleships ordered by Japan under the Ten Year Program of 1896. She was badly damaged by a mine in October 1904 but was repaired in time to play an important part at Tsushima.

Above right: The battered forward 12-inch turret of the *Orel* after her surrender to the Japanese the morning after Tsushima.

Below: Togo's flagship *Mikasa* was the newest Japanese battleship and resembled contemporary British battleships.

considerably or the target had been disabled did the number of hits start to go up. The Japanese system of firing the secondary 6-inch guns to provide ranging splashes was virtually useless.

Another popular misconception was the psychological effect of high-explosive or 'common' shell exploding. Many naval observers claimed that the concussion alone would demoralize an enemy and even advocated using high-explosive in place of armor-piercing shell. Not for another 11 years would it be rediscovered that superficial damage does not sink ships, and only penetration below the waterline or a hit in a magazine can sink a ship quickly. Another point overlooked was that the aggressive tactics used so successfully by the Japanese depended largely on the slow speed of the Russians and their inability to

maneuver in unison. The Russians had other disadvantages too. For the long voyage from Nossi-Bé off Madagascar their ships' ammunition passages and even the decks were piled high with bags of coal. This extra loading was a fire hazard and meant that the shallow armor belt was likely to be completely submerged.

The surrendered Russian ships and those raised in Port Arthur after the capture of the fortress yielded much valuable information, and even the Japanese ships damaged in action showed several shortcomings in design. British technical experts were allowed to inspect some of the ships in Japanese dockyards and the reports drew attention to the need for better pumping arrangements to prevent progressive flooding. Another fault which came to light was the poor quality of the Japanese armor-piercing shell; its Shimose burster was so

sensitive that it detonated the shell as soon as it struck the armor, rather than after penetration. The defect had been spotted earlier and was supposed to have been rectified but many of the Tsushima prizes showed that Japanese shells had burst outside the armor.

The excellent work done by the Japanese armored cruisers obscured the fact that they had been dealing with slow and weakly protected old battleships. A modern armored cruiser was in many ways superior to some older battleships but a less tenable deduction was that such cruisers could take part in a fleet action and trade punches with battleships. At Tsushima Admiral Kamimura's big cruisers had done sterling work but only against Rozhdestvensky's second and third divisions, composed of ships inferior in almost every aspect. It is doubtful if they would have had such a free hand if the Russian battle line had been properly screened by a mixed force of destroyers and cruisers of similar performance.

The main reason for paying too little attention to the lessons of Tsushima was not conceit but the conviction held by the major navies that they were already moving into a new era of technology. Certainly the ships engaged at Tsushima were obsolescent in comparison with the latest battleships on the drawing board in Britain, Germany and the United States and so it was understandable that less attention was paid to their behavior in battle. Many of the new concepts were apparently vindicated by Tsushima and the minor details were dismissed as irrelevant; unfortunately it was these minor details which in the long run caused more trouble than anything else.

The new ideas were basically connected with the need to hit at longer range. It has already been described how improvements in powder had made it possible to increase the length of guns and so improve their ballistic performance, but at the end of the century there was another technological change which went almost unnoticed.

The torpedo had been in service since 1870 but with a range of only 800 yards at 30 knots it held very few terrors at first for battleships on the high seas. However the introduction of the heater system, involving fuel being sprayed into the air vessel and then ignited, pushed up range enormously to 3000 yards and more. This corresponded roughly to contemporary battle ranges, making it necessary for the first time to open fire at much longer ranges. This effect on battle tactics is very rarely mentioned in connection with the sudden increase in the battle ranges expected by all leading navies at the turn of the century and yet it was by far the most important factor. The only drawback was that contemporary gunnery training methods and fire control lagged far behind, and the first attempts to fire at longer ranges were disappointing.

Two captains are connected with the gunnery revolution, the Royal Navy's Percy Scott and the US Navy's William Sims. Scott was a notorious self-publicist but he did manage to push through vital reforms in training which raised the Royal Navy's average performance dramatically. Sims was quieter and devoted himself more to the theory of long-range gunnery, but he was equally influential in having a new generation of ships designed to fight at long range. Given that there were no radar sets and only comparatively crude optical rangefinders there was only one way to range on a distant target: spotting the shell splashes of each salvo. With a battleship firing her four 12-inch guns every 2–3 minutes it was clearly going to be difficult to range on a target if either or both ships were

Left: Seaman practicing at the 6-pounder (57mm) guns aboard the USS *Illinois*. In the days before long-range gunfire these guns would have inflicted casualties and damage on unarmored positions but by 1914 they were not even capable of stopping a destroyer.

moving at high speed. It was also very hard to distinguish the splash made by a 6-inch shell from a 12-inch at great distances since both made very large splashes.

The first solution was to strengthen the main armament by installing an 'intermediate' battery of guns – these ranged from 6.7-inch in German ships to 7-inch and 8-inch in US ships, and finally in 1901 the British followed the trend by putting 9.2-inch in the *King Edward VII* Class. This caused more headaches because the shell splash was even harder to distinguish from a 12-inch and because the blast from such large guns tended to make it hard to man searchlights and light guns nearby. The Japanese went so far as to copy the *King Edward VII*, giving the *Kashima* and *Katori* 10-inch guns but neither ship was ready in time for the war against Russia. The British went one better in the *Lord Nelson* and *Agamemnon*, omitting all the 6-inch guns and increasing the secondary armament to ten 9.2-inch, but the Americans went the whole hog and ordered two ships with eight 12-inch guns and no intermediate guns of any sort.

This was the best solution of all. With four twin turrets a battleship could fire salvos from alternate guns, keeping up a good rate of fire to compensate for the 'rate of change' in relative positions. Four shells in a salvo offered a reasonable chance of hitting, provided the salvo was 'bracketed' around the target. The two ships, the USS *Michigan* and *South Carolina*, were given a most logical arrangement of guns, four twin turrets on the center line, two forward and two aft with No. 2 and No. 3 'superimposed' to fire over No. 1 and

No. 4 respectively. The wheel had turned full circle; after four decades of every possible combination of mountings the basic concept of the old coast-defense ironclads *Prince Albert* and *Royal Sovereign* and the monitor *Roanoke* had reappeared. Quite by chance the 1903 edition of *Jane's Fighting Ships* also published an article by the Italian designer Vittorio Cuniberti, proposing an 'ideal ship for the British Navy' armed with twelve 12-inch guns, protected by 12-inch armor and steaming at 23 knots.

The decision to order the two American ships in 1903, coming on top of the increase in torpedo range and the recommendations of gunnery experts had considerably more influence on the British Admiralty than a sketch design in *Jane's*. Determined not to be left behind, in total secrecy, the British designed a revolutionary new battleship to be called *Dreadnought* and hoped once more to be able to use their superior shipbuilding capacity to regain the lead. The leading influence in this plan was the new First Sea Lord, Sir John Fisher, but paradoxically his motives had little to do with gunnery, torpedo ranges or salvo firing. Fisher wished to enlarge the Royal Navy to meet the challenge from the rapidly growing German navy, and his obsession was economy of operation. If he replaced existing ships with new ships of much greater power he would have to prove to the Treasury that they were cheaper to run and maintain than the older types. For this reason he followed the American lead in restricting the secondary armament to a battery of light guns to fight off torpedo craft but he went much further in

Top right: The Japanese *Kashima* reflected the experience of Tsushima, with secondary guns increased to four single 10-inch.

Center right: The British model for the *Kashima*, HMS *King Edward VII*. She and her sisters were the last William White designs, the end of 10 years of stability in design.

Bottom right: The pugnacious figure of Admiral Fisher (left), the apostle of rapid technological change.

Bottom far right: Italy's first dreadnought, the *Dante Alighieri* on speed trials in March 1914. She reflected Cuniberti's ideas, with four triple 12-inch guns on the center line, and sacrificed protection for higher speed.

Left: HMS *Dominion*, one of eight *King Edward VII* Class built in 1902–05. In addition to the standard pair of 12-inch guns forward and aft she had four single 9.2-inch guns at the corners of the superstructure and ten 6-inch guns in a broadside battery.

adopting the new steam turbine in place of the old reciprocating engines. According to Fisher's calculations 30 of his new *Dreadnought* design could be maintained for a year at the same cost as 29 *Lord Nelson*s, thanks to the simplicity of the armament and the less demanding Parsons turbine.

The argument in favour of the turbine really rested on power output. True, the old reciprocating engines were once run in conditions which made the engine room a 'cross between an inferno and a snipe marsh' but they were now economical and clean thanks to forced lubrication and balanced crankshafts. The real problem was that any major increase of power to give higher speed (say three knots extra) would need nearly 50 percent more power, and such an increase would increase the volume of the machinery enormously. Fortunately the British were in the lead of turbine development, having put the Parsons turbine in four destroyers and a light cruiser. The Cunard line was about to install Parsons turbines in a big transatlantic liner and the RN Engineer-in-Chief was confident that the *Dreadnought* could reach 21 knots, three knots more than the *Lord Nelson*.

To pull together all the diverse strands Fisher appointed a very highly qualified Committee on Designs, including Rear Admiral Prince Louis of Battenberg (Director of Naval Intelligence), Engineer Rear Admiral Sir John Durston (Engineer-in-Chief), Captain John Jellicoe (Director of Naval Ordnance) and Sir Philip Watts (Director of Naval Construction). The Committee's terms of reference were simple: to consider the design of a battleship armed with 12-inch guns and

anti-torpedo boat armament with no intermediate guns, and 21 knots' speed and 'adequate' armor. From the first meeting on 3 January 1905 it was clear that Fisher had already decided what sort of ship he wanted, and there was very little real discussion about alternatives. Yet paradoxically Fisher seems to have had little idea of the implications of long-range gunnery as he still maintained that hits at 6000 yards and more would only be

obtained by firing single shots slowly. Several arrangements of guns were considered in an attempt to get six twin 12-inch turrets but in each case size and cost proved prohibitive or the arrangements were simply not practicable. Finally design 'H' was accepted, and at the insistence of Sir John Durston and Sir Philip Watts reciprocating engines were replaced by turbines at the last minute. When the Committee handed in its report on 22 February it had finalized, in just seven weeks, the most revolutionary warship design since the *Warrior*.

There were surprisingly few mistakes for such a hurriedly designed ship. A subcommitee of Captains Madden, Jellicoe and Jackson failed to appreciate the requirements of long-range gunnery when they sited the boats between the two funnels and hung a boat-handling derrick from the tripod mast leaving the masthead control platform directly over the forefunnel usually to be blanketed by smoke when the ship steamed at full speed. Political pressure on Fisher to keep size down forced Watts to eliminate a strake of 8-inch side armor above the main 11-inch belt, like that provided in the *Lord Nelson*, making the new ship vulnerable to shellfire when the side belt was immersed at maximum draught. Against this *Dreadnought* had several internal innovations and there was much internal weight saving. As it was intended to build the ship as quickly as possible structures were kept simple. The variety of steel sections was kept to the minimum and wherever possible standard-sized plates were used. Finally a large amount of material was prepared in advance and the four 12-inch

turrets already building for the *Lord Nelson* and *Agamemnon* were requisitioned.

With such elaborate attempts to cut down building time it was to be expected that the new battleship would be built in record time, but even so the rate of construction was staggering and it constitutes a record which has never been broken. The keel was laid on Monday 2 October 1905, and only a week later the main deck beams were in place. By the end of December the hull was almost complete and she was launched and christened *Dreadnought* on 10 February 1906. Amid a fanfare of publicity she was credited with completion on 3 October, 366 days after keel-laying but this was a piece of Fisher-inspired nonsense. These were only 'basin trials' and the *Dreadnought* was actually ready to go to sea in December. The results were a great credit to the designers and the builders for the turbines gave virtually no trouble and the hull suffered no ill-effects from the concussion of such a large number of heavy guns. Oddly enough, the principal complaint came from the ordinary sailors for Fisher insisted that the officers should be accomodated forward, instead of aft as was traditional, to be close to the nerve-center of the ship in an emergency. The sailors convinced themselves that they had been put back aft because their officers could not face the terrible vibration from the new-fangled engines! So great was the lower deck dislike of the new arrangement that it had to be dropped in subsequent ships.

The *Dreadnought* put the naval world in a ferment. Although only a logical response to tactical and technical pressures she was

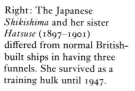

Right: The Japanese *Shikishima* and her sister *Hatsuse* (1897–1901) differed from normal British-built ships in having three funnels. She survived as a training hulk until 1947.

immediately seen as a ship which must be copied. The German *Marineamt* was taken completely unawares by her speed of construction and the massive jump in fighting power, and ordered a halt to battleship construction to give the designers time to absorb the new ideas. In the prevailing mood of Anglo-German tension the newspapers proclaimed that henceforward all comparisons of naval strength must be in 'dreadnoughts' as the 'pre-dreadnoughts' had been made obsolete. This was arrant nonsense, for the *Dreadnought* was simply a more efficient battleship and would in specific circumstances have been at a disadvantage if faced by two smaller ships. But in the age of rampant 'navalism' the British Navy League and the German *Flottenverein* did not bother their heads with such hair-splitting niceties. The arms race between Britain and Germany was given a tremendous fillip, for instead of the Royal Navy's 40 older battleships ranged against 20 German ships of smaller size and weaker armament Britain now had a margin of only one.

Fisher's Committee on Designs had not finished its work with the *Dreadnought* for it also had to consider the design of a new armored cruiser. As soon as the broad outlines of the battleship were settled the Committee started to examine sketch designs but these were like no cruiser yet seen, with 12-inch guns in place of the 9.2-inch carried by the previous *Minotaur* Class. The design chosen bore a strong resemblance to the *Dreadnought* but had nearly double the horsepower to provide 25 knots on a similar displacement. Even so this massive increase in power was only possible because the armor scale was to be the same as the *Minotaur*, 6-inch side armor, and only four twin 12-inch turrets were to be mounted instead of eight. These were the new 'dreadnought armored cruisers' *Invincible*, *Immortalité* and *Raleigh*, (the last two were later launched as *Inflexible* and *Indomitable*), and they were destined to provoke even more controversy than HMS *Dreadnought*. On a displacement of nearly 17,000 tons and armed with eight 12-inch guns there was a natural tendency to equate them with battleships and they were soon unofficially dubbed 'battleship cruisers' and finally in 1912 'battlecruisers.'

The problem was that their proper role was never clearly worked out. Internally they showed very little difference from the *Minotaur* Class, and they were in no sense fast battleships. Fisher had been greatly impressed by the success of the Japanese armored cruisers at Tsushima and wanted ships which could reconnoiter for the battle fleet, pushing past any small cruisers which tried to stand in their way. For this purpose the 12-inch guns made sense as they would allow the big cruiser to get close enough to the enemy battle line even to count numbers and estimate course and speed. Such a ship could not be dealt with easily by small ships and would in the closing stages of a fleet action be able to swoop on crippled battleships as well. So far so good, but in the second role of the armored cruiser, protection of commerce, the 12-inch guns were far too big and slow-firing to be of much use against fast-moving targets. The new type was therefore a big and expensive solution to that problem. Fisher maintained that the *Invincible* was fast enough to catch anything smaller or escape from anything more powerful, but this presupposed that no other navy had similar ships.

The real weakness of the battlecruiser was that its heavy armament lent it a spurious 'capital' rank, and an admiral would always be tempted to use it to reinforce his battleships. In fact within a year or two of their introduction the term 'capital ship' was introduced to cover both battleships and battlecruisers. The wide disparity between the Russians and the Japanese at Tsushima was overlooked, and nobody seems to have thought of what might have happened to Admiral Kamimura's armored cruisers if they had been opposed by a better-trained enemy. A ship with a 6-inch armor belt would be vulnerable if she attempted to fight a modern battleship on equal terms, but Fisher airily dismissed this question by saying that 'Speed is Armor.' In one sense that dictum was perfectly valid, but only if the battlecruiser used her superior speed to stay out of trouble. A fast battleship would have been the ideal means of gaining a tactical advantage in battle but a 25-knot ship protected with 12-inch armor would have had to be 50 percent bigger, and this was not politically or financially acceptable.

The Germans replied to both the *Dreadnought* and the *Invincible*, the former with the 20,000-ton *Nassau* Class and the latter with the

Above: The Russian *Peresviet* was sunk by howitzer shells at Port Arthur in December 1904 but was salved a few months later and repaired. She is shown here later still, in Japanese service when she was named *Sagami*.

19,000-ton *Von der Tann*. The battleships were not an outstanding design, quite well protected but carrying their six twin 11-inch guns in a cramped arrangement of center-line and wing turrets, whereas the battlecruiser was a major improvement over the *Invincible*, with good protection and layout. The German battlecruiser designers got closer to the ideal of the fast battleship because they were prepared to sacrifice some speed and gunpower to provide more armor; the 11-inch gun was retained to allow 8-inch armor and the extra tonnage was used to provide better protection against torpedoes. The *Von der Tann* makes an interesting comparison with the *Indefatigable*, a repeat of the *Invincible* with no major improvements and all the faults of the prototype:

Comparison of Von der Tann and Indefatigable

Weights (tons)	Von der Tann	Indefatigable
Hull	6004 (31.5%)	7000 (37.4%)
Machinery	3034 (15.9%)	3655 (19.5%)
Armor and Protection	5693 (29.8%)	3735 (19.9%)
Armament (including turret shields)	2604 (13.7%)	2580 (13.8%)

Right: SMS *Von der Tann*,
first of the German
battlecruisers, lying in the
Jade River in 1918.

Right: The battlecruiser
HMAS *Australia* fitting out
in 1912. The starboard
12-inch turret has not yet
been installed.

Center right: The
Australia's massive ram bow
was for esthetic effect and
not for use against other
ships. The *Australia* was a
sister ship of the
Indefatigable.

Although only a rough comparison it shows the penalty paid by the British ship in extra weight devoted to machinery. Weight of armament was, however, roughly equal as the British had more experience in keeping down turret weights, and did not pay any appreciable penalty for having four 12-inch turrets instead of four 11-inch.

The British relied on their superior shipbuilding capacity to increase their lead over the Germans, and deliberately made no major alterations to the succeeding classes of dreadnoughts to avoid delays. In this way they had seven dreadnoughts of nearly identical design in service by early 1910, *Dreadnought* herself, three *Bellerophon* Class and three *St Vincent* Class, as well as three *Invincible* Class battle-cruisers. In the same period the German navy commissioned four *Nassau* Class and the *Von*

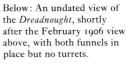

Left: A dated sequence of *Dreadnought* under construction. Top row (from left): The bare slip, 1 October 1905; 28 October, with the sloped armored deck being laid; after launch, in April 1906 the machinery and armament is being installed in the hull. Bottom row: On 2 October 1905, the keel and some frames are already in position; the launch on 10 February 1906; 11 August 1906, nearly complete with masts, funnels and turrets in place.

Below: An undated view of the *Dreadnought*, shortly after the February 1906 view above, with both funnels in place but no turrets.

der Tann was still completing. Even the next three British dreadnoughts, the *Neptune* and the two *Colossus* Class and the three *Indefatigable* Class battlecruisers were very similar to their prototypes apart from a modified layout of guns to improve their arcs of fire. In their second class of dreadnoughts, the *Helgoland* Class, the Germans decided to adopt a 12-inch gun to match the British gunpower but the British had already decided to put 13.5-inch guns into the *Orion* Class of 1909.

The race was now an open one, with each class of ship intended to match the latest on the other side of the North Sea. It was useless for the diplomats to try to negotiate any 'holiday' to slow down the tempo of this arms race, for the answer was always 'national survival is at stake.' The cost of *Dreadnought*, £1,800,000 ($9,000,000 then), sounds absurdly cheap today, but in real terms it represents at least 100 times that figure. The reasons for this apparent willingness of both nations to rush

Right: The stokehold of this British dreadnought gives a misleading impression of tranquillity. Normally it would be crowded with stokers bringing coal to the fireboxes.

Below: The *Thüringen*, one of the first 12-inch gunned German dreadnoughts. She and her sisters of the *Helgoland* Class were built in 1908–12.

toward Armageddon are subtle and complex and have no place in this book, but it must be remembered that the outward calm of the Edwardian Age masked currents of unrest that would only come to light during and after World War I. For example, right up to the outbreak of war men were being laid off in German shipyards; when the *Dreadnought* was built at Portsmouth it was recorded that the workmen did not like working 69 hours a week but 'with so many being discharged it was not difficult to persuade them to accept it.'

The building of dreadnoughts can be equated with the space program in the 1960s, in the sense that it fuelled a crucial area of economic growth and demanded the very best of contemporary technology, but was also inextricably linked with national prestige. A battleship provided work (nearly 3000 out of Portsmouth's labor force of 8000 worked on

by 1908 the small ineffective coast-defense ships had been replaced by ocean-going battleships clearly capable of fighting their opposite numbers in the Royal Navy, and the British did not like the implications. In 1908 Tirpitz used a loophole in the current Navy Law to make a major increase in the strength of the German navy; a provision for replacing eight *grossekreuzer* (large or armored cruisers) was interpreted as permitting the building of eight battlecruisers. Although the precedent had been established when Fisher referred to the *Invincible*s as dreadnought armored cruisers the German move came at a time when German naval expansion was under particularly close scrutiny in Britain. The Liberal government was under strong pressure from its Radical wing to reduce 'bloated armaments' but equally torn by fear of losing popular support if it allowed the Royal Navy to be overtaken by the German navy as a result of unilateral disarmament. There was also the fear of industrial unrest if the tempo of construction should slacken, one more worrying limitation for the Liberals. In Germany too, the situation was not straightforward as the military hierarchy was constantly worried by the threat of a Socialist majority in the Reichstag. People outside Germany tended to believe the posturings of the Kaiser as Supreme War Lord but in fact he was no more than a figurehead and the figurehead of an insecure military hierarchy at that.

Dreadnought) in the shipyards but also in the steelworks and gun foundries of the nation. Heavy engineering was still the motor force of industrial expansion and the battleship's needs for more resistant armor plate and better shells, guns and machinery all stretched various sciences to the limit. The limit to which armor plate could be hardened without losing other qualities was reached in the early years of the twentieth century but metallurgists remained fully occupied with questions about lighter structural steels, shell casings and heavy forgings for guns and machinery. The chemists had already transformed long-range gunnery by producing nitro-cellulose propellants but research continued into shell bursters and safety in handling.

The last chance to effect a reconciliation between German and British interests probably came in 1908. The architect of German naval expansion, Admiral Tirpitz, had successfully fought for his Navy Laws back in the 1890s, legislation which provided for regular replacement of obsolete tonnage instead of the annual haggling between the Navy and Treasury and Parliament, such as went on in Britain. In this Tirpitz was only doing his duty toward his country, to prevent later governments from hamstringing a long-term program by neglect or sudden whim. But

The *Dreadnought* did not cause World War I any more than Mahan's *Influence of Sea Power on History*, but she and her successors were an integral part of the last decade of the old order that passed away in 1914. They were a symptom of the fever in the veins of Europe, not the fever itself, but there can be no doubt that the endless comparisons of strength and the forecasts of future trends inflamed public opinion and played on deep-rooted fears. The British slogan, 'We Want Eight and We Won't Wait,' expressed the fears and hopes of xenophobes in all the major maritime nations.

4. RETURN OF THE BIG GUN

The speed with which the *Dreadnought* and her successors had been built made it more than likely that some flaws in design would come to light. Although *Dreadnought* herself was a technical success in that all systems worked well and considerable weight had been saved without weakening the structure, she had two obvious faults. One was the layout of the 12-inch turrets, with two wing turrets abreast of the bridge, and the other was the position of the tripod mast between the funnels. The former was intended to improve end-on fire but having the turrets so close to the bridge-work raised formidable problems of blast, while the latter made absolutely certain that funnel smoke would blanket the fire-control position.

The Germans had reacted as quickly as they could by building the *Nassau* Class, but as already mentioned, the designers adopted the clumsy hexagon arrangement, with two pairs of turrets on the beam, one turret forward and one aft. The reason was simply to be able to reply to ten 12-inch guns with twelve 11-inch, but as in the *Dreadnought* the layout still only permitted a broadside of eight guns. In the next class, the *Ostfrieslands*, the drawbacks became even more apparent as the bigger diameter of the 12-inch SKL/50 gun mounting encroached severely on the anti-torpedo protection. In general the Germans devoted more care to underwater protection than their British counterparts, relying on a heavy torpedo bulkhead set about 13 feet inboard from the hull. Machinery spaces were well subdivided; the *Nassau* Class, for example, carried their 12 boilers in six separate compartments and the *Von der Tann* had nine boiler rooms. The pumping and flooding arrangements were generally good in the central portion of the ship (between the forward and after barbettes) but particularly in the forward section German designs were prone to rapid flooding. A source of weakness was the submerged torpedo-tube compartment, which was difficult to keep watertight.

A legend has grown up crediting the German dreadnoughts with unusually elaborate magazine precautions, but in fact they were on the whole rather primitive by some other navies' standards. The shells and cordite (nitro-cellulose) charges were stowed together, not separately as in British ships, but because the main cordite charge was kept in a heavy brass case in the magazine, and most of all because the black powder igniter at the bottom end of that charge did not come into contact with the other two charges from the time it left the magazine until it was fired from the gun, there was less risk of accidental explosion. In British ships the igniter was sewn in a small pocket of silk at both ends of all charges, and the very fine high-quality gunpowder tended to seep through the silk, eventually coating the outer covering of the cartridges. If charges were placed end-to-end during the loading sequence a flame at one end could pass very quickly down the line of charges.

Cordite, made of gelatinized nitro-cellulose, had been in service since the early 1890s, and the Cordite MD used by the Royal Navy in 1914 had been introduced in 1901. Ordnance experts had known for many years that if it was not correctly made from pure raw materials it had a nasty habit of exploding without warning. On 12 March 1907 the French battleship *Iéna* blew up in the dock at Toulon when a secondary magazine overheated and on 25 September 1911 an even worse explosion destroyed the *Liberté*, demonstrating the risks. Not for many years would it be discovered that a small quantity of iron pyrites in the atmosphere could result in serious instability. Another source of instability was the solvent, usually petroleum jelly, and in its RPC/12 propellant the German navy managed to replace it with a mixture of nitroglycerine and a compound known as Centralite. This resulted in a very stable propellant which would burn if exposed to flame or explosion but would never flash

These problems were not common knowledge, and indeed many naval officers remained permanently in the dark about them. In 1908 the Royal Navy tested its current Armour Piercing (AP) shell against the old turret ship *Edinburgh*, which had been fitted with sections of plating representing the protection of the latest dreadnoughts. To the consternation of the Director of Naval Ordnance, a shell striking the plate at an angle much off the perpendicular was not likely to penetrate. The use of a sensitive burster made of Lyddite and slightly overhardened nose caps meant that the shell was likely to break up on impact and the burster was then likely to detonate prematurely. The shell would thus expend most of its energy outside the enemy battleship's armor instead of bursting deep inside its vitals.

The Director of Naval Ordnance instructed the manufacturers to look into both problems but little or nothing seems to have been done about them. Trinitrotoluene (TNT) made a much better burster but it was more expensive to produce than the standard British explosive Lyddite, and the Treasury was reluctant to authorize any such expenditure. The problem of brittle nose caps was one which had occurred in the 1870s, when Palliser 'chilled' shot had sometimes been cooled too rapidly during the forging process. The shell manufacturers were allowed to submit shells for batch-testing, and if the normal two percent of the batch supplied were found to be faulty when tested, the manufacturer had the option of submitting another two percent or withdrawing the whole batch of shells. Naturally the manufacturers always chose the former course, in the hope that the next two percent would pass the test. Unhappily this procedure could conceal a very high percentage of failure, and there was very little incentive for industry to improve standards. To make matters worse the Royal Navy in 1908 still had no proper general staff, and the results of the *Edinburgh* trials were never followed through.

Despite these hidden weaknesses the later British dreadnoughts were powerful ships. After the initial batch of copies of *Dreadnought* herself, three modified ships were built, the *Neptune,* *Colossus* and *Hercules*, with an altered layout of guns. The original layout meant a broadside of only eight guns, and these ships were given a superimposed turret aft and had the midships wing turrets staggered. In theory this arrangement had a lot to recommend it but actually it caused severe blast damage to the superstructure and cross-deck firing tended to strain the hull. Finally the Admiralty decided that however ingenious broadside positions might be, the best solution was to put all turrets on the center line. This had been achieved in the USS *Michigan* and *South Carolina* but the principle had first been demonstrated in the 1860s by the small turret ships *Prince Albert* and *Royal Sovereign*.

It was also realized that the 12-inch gun was reaching the limit of its development. Attempts to boost range by keeping the shell weight down and pushing up muzzle velocity with a bigger charge were causing a loss of accuracy

Left: The *Neptune* was the first British break away from the original *Dreadnought* layout, with staggered wing turrets and 'X' turret superimposed, in order to provide a full 10-gun broadside.

Below: The *New York* in Hampton Roads on 10 December 1916. She and her sister *Texas* introduced the 14-inch gun to the US Navy and they served through World War II.

Below: HMS *Monarch*, one of the so-called 'super-dreadnoughts' with 13.5-inch guns. The floatplane is almost certainly a fake.

and excessive barrel wear. Longer barrels tended to whip when fired and the light shell tended to 'wander' toward the end of its flight. The Royal Navy seems to have been the first to spot this problem, for in 1909 the Admiralty ordered a new class of ships armed with ten 13.5-inch guns. The four *Orion* Class were impressive looking ships with two turrets forward, one amidships and two aft, and although underwater protection was poor and the pernicious habit of putting the tripod mast between the funnels was continued, the new gun was an outstanding success. Not only did

it shoot a heavier shell farther but with a lower muzzle velocity there was much less barrel wear. In the next mark of 13.5-inch the weight of shell was pushed up from 1250 pounds to 1400 pounds, making it an even better gun for long-range shooting.

Across the Atlantic the US Navy was keeping pace with European developments, but under considerably less pressure the General Board was able to take a longer view of the future requirements for dreadnoughts. The fleet was not happy with the first dreadnoughts, the *Michigan* and *South Carolina*, nor with the *Delaware* and *North Dakota*, which all had a more efficient arrangement of armament than the *Dreadnought* but nothing like as satisfactory a propulsion system. Only the *North Dakota* was given turbines (and these had to be replaced five years later) whereas her sister had

reciprocating engines. The next class, the two *Florida*s were very similar, but in the *Wyoming* and *Arkansas* an attempt was made to improve gunpower by adding a sixth twin 12-inch turret. As this was achieved without resorting to wing turrets the designers either had to accept a very long and much heavier hull or choose a very cramped arrangement.

In the next class, the *Texas* and *New York*, there was no question of adding another turret, and to increase gunpower while reducing the number of turrets to five, it was necessary to adopt a heavier gun. As the British had already announced that the *Orion* Class would have 13.5-inch guns the General Board ordered a 14-inch gun firing a 1400-pound shell, but the fleet was still unhappy with the ships when they appeared for the five-turret arrangement was felt to be awkward and the thin armored deck was recognized to be inadequate against long-range plunging fire. By May 1910 the Bureau of Construction and Repair (BuC&R) had prepared a sketch design with four triple 14-inch turrets, capable of 23 knots and protected by 9-inch to 11-inch armor. The problem was how to protect such a ship against severe damage from heavy shell hits; at 15,000 yards or less the latest 12-inch/50 caliber would penetrate 11-inch armor and the new 14-inch gun would do this even more easily.

In considering the new design there were other factors to be taken into account. The first four dreadnoughts were now at sea, and reports from them showed that the magazine of the midships turret suffered from over-heating because it was inescapably near the boilers. On the credit side there was an enthusiastic report from the *Delaware* about her experimental oil-fired boilers. Not only was there more thermal efficiency in oil and so more range, but a number of other improve-

Above: A *New York* Class battleship, part of the 6th Battle Squadron at the surrender of the High Seas Fleet in November 1918.

Top: The USS *Delaware* and *North Dakota* (1907–10) followed the *South Carolina* Class. *Delaware* is seen here at Rosyth in 1918.

ments were also made. The boiler rooms were smaller and so required less extensive armor protection and they could be served by only 24 rather than 112 men. There was no need to reduce speed periodically to clean clinker out of the boilers and refuelling was also quicker and cleaner. Coal-fired systems also took longer to raise steam from cold.

However, the argument was not quite so simple, and BuC&R pointed out that some additional internal armor would have to be provided to compensate for the loss of protection afforded by water-line coal bunkers. Furthermore with the oil stowed along the length of the ship below the water-line she would be less steady and more weight would be carried at the ends of the hull. But the advantages were overwhelming, particularly as the United States controlled very large oil reserves, and in November 1910 the momentous decision was taken that oil would henceforth be the only fuel for US battleships.

As in most peacetime designs the General Board was hampered by restrictions on size and cost, particularly in this instance as Congress had authorized a cost for hull and machinery of exactly $6,000,000. The General Board estimated this as 27,000 tons, assuming that the May 1910 Characteristics were followed. As these parameters involved a ship 588 feet long and capable of docking in only four places between Pearl Harbor and New York (none of these drydocks was complete) the design could hardly be called satisfactory. Finally the Board decided to cut the main armament to 10 guns by substituting twins for the superimposed triple 14-inch and cut speed to 20.5 knots, with endurance of only 8000 miles.

As the Board did not want to sacrifice protection the BuC&R designers were forced to accept a radical departure from previous ideas, an 'all or nothing' scheme. This finally evolved into a heavy, deep belt of 13.5-inch side armor covering the machinery and magazines, with a heavy deck of 3-inch armor. Its essentials were simple: over a certain distance the plunging shells would go over the top of the belt, and so increased protection from long-range gunfire should be confined to the deck only. From this stemmed the concept of the 'immunity zone,' a minimum range at which shells would be stopped by the side armor and a maximum range at which plunging shells would be caught by the deck armor. It was recognized that some shells might penetrate the deck, but as the shells would be travelling at much less than their initial velocity all that was needed was a 'splinter deck' below the armor to catch any fragments which came through. So successful was this concept that it remained valid as the basis of US battleship design for another 30 years, and was ultimately copied elsewhere.

An important part of the weight saving was the adoption of a triple turret, for it allowed the side and deck armor to be increased to the point where they were significantly better than in previous classes. To keep weight down all three guns were kept in a single sleeve, with only two trunnions supporting all three barrels. As the US Navy was using a 'follow the pointer' system of fire control it would have been very difficult to position the central pointer if the guns had been elevated independently. Although fears were expressed that a single hit would knock out 30 percent of the ship's guns and the designers also worried that three large apertures in the face armor might weaken the turret's structure, the order was authorized in January 1911. Three months later the ships themselves, the *Nevada* and *Oklahoma* were authorized. By the time they came into service in the spring of 1916 they

Below: The USS *Nevada* escorting the liner carrying President Wilson to Brest in December 1918.

were overshadowed by later designs but the *Nevada*s stand out as the biggest step forward in battleship design since the *Dreadnought*. In their stark simplicity, both in conception and in appearance, they have more in common with the old *Devastation* and *Thunderer* than their contemporaries.

The Japanese felt compelled to match the American 14-inch gunned ships and when the Diet voted funds to build four large battle-cruisers under the 1910–11 Program the Imperial Navy asked Vickers to arm them with 14-inch rather than the 13.5-inch. The British armaments industry was so huge that Vickers were able to produce a 14-inch without stretching their resources, although there was little to choose between the two guns. The design was entrusted to Sir George Thurston, the Chief Naval Architect of Vickers, who followed the main features of the new *Lion* Class and the *Queen Mary*, the Royal Navy's first 13.5-inch gunned battlecruisers. Thurston was, however, able to incorporate certain improvements already approved for the *Queen Mary*'s significantly modified sister *Tiger*, principally a rearrangement of the guns and boilers to avoid a particularly inconvenient layout in the first three ships.

The first Japanese battlecruiser was named *Kongo* and was launched in May 1912 and delivered the following year. Her three sisters *Hiei*, *Kirishima* and *Haruna* were built in Japan with 14-inch turrets imported from Great Britain, and when the last ship was completed in 1915 Japan had a squadron theoretically superior to anything the US Pacific Squadron could muster. At 27,500 tons, armed with eight 14-inch and steaming at 28 knots they appeared to be able to hold their own against anything, but their 8-inch side

armor would not have kept out heavy shells at likely battle ranges. Despite this weakness they were proof of Japan's growing industrial might; not only was the *Kongo* the last major warship built outside Japan but the 14-inch guns of the *Haruna* were the last gun-mountings to be imported.

The German navy was now outgunned badly and no amount of argument about the merits of light shells and high muzzle velocity could disguise the fact that a bigger gun fired a bigger shell further. But the real problem lay in the fact that Krupp's factory at Essen could not turn out guns very quickly – a penalty paid for excellence. When the *Derfflinger* and *Lützow* were laid down in 1912 in reply to the British *Lion* and *Princess Royal* they were only armed with 12-inch guns, although the SKL/50 with its 893 pound shell was a great improvement over the 11-inch SKL/50 in the previous battlecruiser, SMS *Seydlitz*. The same armament was used for the *Hindenburg*, a half-sister ordered a year later. The *Marineamt* was not prepared to sacrifice protection in favor of

Above: The former flagship of the High Seas Fleet, SMS *Grosser Kurfürst* at the November 1918 surrender, as seen from a British airship.

Below: The Japanese battlecruiser *Hiei* on completion in August 1914. The forefunnel was raised during fitting out to keep smoke away from the bridgework.

gunpower, and as the *Derfflinger* already displaced 26,000 tons there was considerable political opposition to any increase in size and cost. The German navy paid a high price for the excellence of its ships: the battleship *König*, for example, cost £2,400,000 whereas the *Lion* cost some 25 percent less, and a vocal Socialist minority in the Reichstag was growing restive at the sums expended on the navy.

The British response to the growth in gun-caliber was quick and decisive. Late in 1911 the Elswick Ordnance Company was asked by the Admiralty to investigate the feasibility of building a 15-inch gun, the intention being to leapfrog over the competition and ensure a useful margin of superiority for some years to come. The response was encouraging; the manufacturers were confident that they could handle the larger forgings and that there were no risks inherent in such a jump in size. At first the intention was merely to scale up the existing *Iron Duke* Class and mount five twin 15-inch turrets in the same disposition, two forward, one amidships and two aft but the War College advised that it might be valuable

to have a 'Fast Division' of battleships capable of steaming at 25 knots. This marks the beginning of the end of the battlecruiser idea, for the Fast Division was required to carry the same scale of armor as a battleship but be capable of maneuvering independently of the main battle line, either to force an enemy battle line to change course or to finish off damaged battleships.

It is not known to what extent the British Admiralty was influenced by the American decision to use oil fuel in the *Nevada* Class, but exactly a year after the American ships were ordered the British also decided to convert entirely to oil fuel. Although not possessing the vast US oilfields the British controlled the Middle East and to guarantee access to oil the British Government bought a majority interest in the Iranian oilfields. The arguments were exactly the same as those which had influenced the US Navy's General Board, and as with the *Nevada*, the new *Queen Elizabeth* Class needed oil fuel to achieve the aims of the designers. By dropping the midships turret, which was in any case badly sited, space could be found for

Right: HMS *Iron Duke* on trials early in 1914, still with torpedo nets and lacking a complete foretop.

Below: The *Iron Duke* and *Marlborough* and the *King George V* Class arriving in Spithead in July 1914 for the trial mobilization which fortuitously put the Royal Navy on a war footing.

nearly double the horsepower without sacrificing the weight of protection. The result was the first true fast battleship, for the *Queen Elizabeth* was protected by a 13-inch belt and had a designed speed of 25 knots. In fact the amount of extra weight worked in during construction pushed normal displacement up by nearly 2000 tons and cut speed by a knot, but they were still the finest British ships built to date, and because British shipyards could build so quickly the lead ship was ready just after the outbreak of war in August 1914.

Nor did the armaments industry fall behind the shipbuilders. With the likelihood of war with Germany becoming greater each year, the Admiralty wanted no delays in getting the new Fast Division into service. The main risk of delay was the time taken to evaluate a radically new gun, but in this instance the experience with the 13.5-inch and 14-inch at Elswick was sufficient for the makers to guarantee that there would be no problems. To avoid any delay in building the ships the entire outfit of guns for eight ships plus a reserve of spares (some 50 barrels) were ordered, with one barrel to be completed early enough to permit some proof-firing before the ships were ready. The gamble paid off and the 15-inch/42 caliber Mk I proved an outstanding success. It was still in service 40 years later and the mounting was widely regarded as the most efficient and trouble-free built for any navy. It was the outcome of a steady process of development going back to the end of the Victorian Age, when the prototype of the hydraulic all-round-loading turret was perfected.

The weak points of the *Queen Elizabeth* design were, however, as marked as the good ones. The magnificent 15-inch guns had the same faulty shells as earlier marks of guns and their cordite propellant was just as unstable. Neither was the scale of armoring as good as that in the *Nevada* or the latest German dreadnoughts; the narrow 13-inch belt was almost totally immersed at deep load, leaving strakes of 6-inch armor to keep out enemy shells. It is a sad commentary on the design of

these ships that they were not as well protected as contemporary German battlecruisers. The *Derfflinger* on a similar displacement (26,000 tons) was protected by a deeper and more extensive belt of 12-inch armor and much better underwater protection, and in addition she could steam at 26.5 knots. Of course the British ship had an advantage of sheer weight of metal and was more seaworthy, but her defective shells and dangerous propellant

Above: British sailors and marines 'rubbernecking' aboard the salved battleship *Baden*. She was the only major unit of the High Seas Fleet not successfully scuttled at Scapa Flow in 1919.

Top: The German *Baden* in 1917–18. The circles on the turret tops were to avoid attack from German aircraft.

completely offset her advantage in gunpower.

The Germans were finally shaken in their faith in the 12-inch gun, and to respond to the clear advantages of the 13.5-inch and 14-inch guns in other navies decided to produce two new calibers, a 13.8-inch L/45 for the battle-cruisers to be started in 1915 and a 15-inch L/45 for the 1913 battleships. By a strange quirk of fate, British and German designs, which had been on diverging paths for some years, came together to a remarkable degree in these battleships. Known as the *Bayern* Class, they were to displace 28,000 tons and be armed with eight 15-inch in four twin turrets. As the last German capital ships designed before World War I they were the high point of a process which had started with the *Nassau* and the *Von der Tann*, but what was more re-markable was their marked similarity to the British *Revenge* Class, which followed the

Queen Elizabeth into service in 1916. The lay-out was similar, as were speed and armor thicknesses, and it was as if a common specification had been given to each design team. In fact the Germans had no knowledge of the British 15-inch, which for security was referred to merely as the '14-inch A' during the design stage.

The other countries in the race to build dreadnoughts came a long way behind the Big Three. Austria-Hungary built nominally im-posing ships, the *Tegetthoff* Class with four triple 12-inch turrets to oppose Italian ships armed with ten 12-inch but they were all too lightly protected and had low endurance. France, having once looked like overtaking Great Britain, had seen her navy slide rapidly to the second rank as a result of maladmini-stration. When the *Dreadnought* appeared in 1906 there was no longer any threat from the

Top right: The French dreadnought *Provence* off Dolma Bagche, near Istanbul in November 1919. French and British heavy units moved into the Black Sea after the armistice to support the White Russian armies against the Bolsheviks.

Center right: The *Paris* on full-power trials shortly before the outbreak of war in August 1914.

Below right: The 'inter-mediate' dreadnought *Danton* was armed with four 12-inch and twelve 9.4-inch guns.

Above: The Austro-Hungarian dreadnought *Prinz Eugen* ready for launching at Trieste in November 1912.

Above center: The *Viribus Unitis* firing a broadside.

Below: The French dread-nought *Jean Bart* had six twin 12-inch gun turrets, two forward, two aft and two amidships. She survived a torpedo hit in December 1914.

Royal Navy thanks to the *Entente* but even if there had been France lacked the industrial capacity to make any serious reply. The six *Danton* Class, although elevated to the curious hybrid rank of 'semi-dreadnoughts,' were hardly a match for the latest British pre-dreadnoughts, and yet they were not completed until 1911. Even the first proper dreadnoughts, the *Courbet* Class built in 1911–14, were slow and lightly armored. French experience with Parsons turbines was not happy and there was no move to oil fuel, apart from continuing its use as a supplement to coal.

The second group of French dreadnoughts, the *Provence* Class, were considerably better but they still suffered from having rather too heavy an armament on a small hull and so could only be lightly armored. On such a relatively small hull it was necessary to mount the forward 13.4-inch guns as far forward as possible and as they also had armor up to the bow they buried their forecastles in a seaway. This hampered their seakeeping to such an extent that eventually 10 meters of the armor belt had to be removed. Under the Naval Law of March 1912 it was hoped to catch up some of the lost ground by building five *Normandie* Class, 25,000-tonners armed with a unique arrangement of three quadruple 13.4-inch gun

Above: The Italian dreadnought *Leonardo da Vinci* passing the swing bridge in Taranto. She was destroyed by an internal explosion in August 1916.

Below: The British predreadnought battleship *Redoubtable* fires a salvo at a shore target early in World War I.

turrets, and had these been completed they would have been the most unusual looking ships afloat. But as we shall see they were overtaken by events.

The Russians were making a very slow recovery from their losses at Tsushima but they had every intention of possessing a front-rank navy once more, however long it took. After the war they revised the design of two battleships already under construction and completed them at a majestically slow pace by 1910–11. The first dreadnoughts were laid down in 1909, four for the Baltic Fleet, followed by three similar ships for the Black Sea Fleet in 1911. The specifications were originally put out to foreign shipyards for tenders and the contract was just about to be awarded to Blohm and Voss of Hamburg when the Imperial Government intervened with a stipulation that all the

ships should be built in Russian yards. The design staff in St Petersburg then revised the design to incorporate what they perceived as the best features of all the foreign designs, particularly an Italian one reflecting the concepts of General Cuniberti.

The ships which resulted were startling, with a long flush hull and four triple 12-inch turrets on the center line, but although they looked good on paper they were weakly protected, lacked structural strength and were credited with ventilation bordering on the unsanitary. The three Black Sea ships were about 1.5 knots slower but slightly better protected than the Baltic ships, but all seven were equipped with excellent long-base rangefinders which, combined with higher elevation to their guns than was normal in other navies, enabled them to shoot accurately at long range.

The problem of hitting at long range was the one which received least attention, despite proving more intractable than any other. Guns, armor and machinery had all improved beyond anything conceived at the turn of the century and yet long-range gunnery was still a very chancy affair. Shooting had improved from the low level common to all navies in the 1890s to a new peak under such men as Scott in the RN and Sims in the USN but at ranges of 10,000 yards and more there was little likelihood of more than two shells out of every hundred hitting. Many people, not least Lord Fisher, assumed that a ship only had to fire her guns rapidly to ensure hits.

The Germans had the benefit of good optics from the Jena works of Carl Zeiss, and a typical dreadnought such as the *König* had seven *Basis Gerät* stereoscopic rangefinders, one on each of the two conning towers and in the roof of each 12-inch turret. On the foremast was a spotting top from which fall of shot could be observed, and these observations were relayed to the control tower, which was inside the main conning tower. The unique features of German fire control were the use of stereoscopic instruments and the 'ladder' method of finding the range. The operator turned a knob until his *Wandermark* (a cross or arrow) appeared in the eyepiece directly over the object. As soon as the gunnery officer in the control tower had the average range from the various rangefinders he ordered a salvo (usually one gun from each turret but in range-finding three guns) fired: one shell on the range given, the second up by a fixed distance and the third down by the same distance. This was a 'ladder' which the remaining gunlayers could follow if the target ship changed course; during all this each gunlayer kept his sighting telescope trained on the target to compensate for the roll and pitch of his own ship.

The British had never liked the stereoscopic rangefinder and preferred the 'Incidence' type in which the range taker had to reconcile two halves of the image split horizontally. Both types had their advantages; the German allowed ranging on ill-defined objects such as plumes of smoke but required intense concentration which could be distracted by excessive vibration and noise, while the British needed a more definite image but was less demanding to operate over a long period. Where the British were considerably inferior was in relying on a short-base rangefinder for too long. As late as 1911 the Admiralty specified only a 9-foot Barr & Stroud rangefinder, and the *Queen Elizabeth* Class of 1912 were the first to have a 15-foot rangefinder.

To correct the deficiencies in their gunnery the Royal Navy revived an idea which had been first thought of in the 1880s, central director-firing. It was the brainchild of Rear Admiral Percy Scott, and comprised a central gun control station (the 'Director') high up on the foremast, from which a gunnery control officer could fire all the guns when in *his judgement* they were on target. Being high up he was

Bottom: The *Duilio* and her sister were armed with 13 12-inch guns in triple and twin turrets, and could steam at 22 knots.

Below: The Japanese *Tsukuba* and her sister *Ikoma* (1905–08) were hybrid battleship/cruisers with the hull of an armored cruiser and the armament of a predreadnought.

clear of spray and smoke and had a better view of the target, and provided he had the range accurately he was likely to be far more accurate than individual gunlayers in the turrets below. The first trial of Scott's Director System took place when the new dreadnought *Neptune* was fitted with it, but in November 1912 there was a competitive trial between the *Orion*, currently the crack gunnery ship of the Home Fleet and her sister *Thunderer*, specially fitted with a Director.

The *Thunderer* beat her sister so decisively in the trials that an immediate order went out to retrofit the system to all the dreadnoughts as they came in for lengthy overhaul. It was a timely decision for the clouds of war were gathering. Director-firing involved great lengths of electrical cable for transmitting data to the director from the spotters, down to the transmitting station deep in the bowels of the ship and then passing the firing signal back to the guns. The British were fortunate that their dockyards were able to cope with such a vast workload, but even so some of the older dreadnoughts were still being fitted with director-firing equipment when World War I was some months old.

The acceptance of director-firing was comparatively painless but when an inventor called

Arthur Pollen produced an 'Aim Corrector' for calculating mathematically the solution of all the variables, course, speed, range and others, he found it much harder to get a hearing from the Admiralty. As early as September 1906 Fisher had written to the First Lord of the Admiralty to say that the Royal Navy must acquire the rights to Pollen's invention and yet in 1913 it was still not adopted. The reasons are as much to do with the people chosen by Fisher to run the navy as anything, but they do not make pleasant reading. First supervision of the trials was entrusted in turn to a pair of torpedo specialists and second to a gunnery officer who had invented a rival to Pollen's gear. Under such circumstances it was unlikely that Pollen's 'Range Clock,' the forerunner of modern fire-control computers, would receive a fair hearing, but it was reprehensible that features of his patented design should reemerge very shortly afterward in Commander Dreyer's 'Fire Control Table.' Recognition was withheld from Pollen until 1925, when after a stiff legal action he was awarded compensation, but fortunately for the Royal Navy the essentials of his invention were grasped and put to work in time to improve the Home Fleet's gunnery for the headlong clash with Tirpitz' creation, the High Seas Fleet.

Top left: The Japanese dreadnought *Settsu* on trials in 1912. She was armed with six twin 12-inch turrets but unlike British and American contemporaries had a heavy secondary battery.

Center left: The French *Vergniaud* belching smoke from her 26 coal-fired boilers.

Left: Predreadnoughts of the High Seas Fleet at sea on exercises.

Top: The German predreadnought *Wittelsbach* (1898–1902) and her four sisters were built under the 1898 Navy Law.

Above: The launch of the *Imperator Pavel I* on 7 September 1907 at St Petersburg.

5. THE THUNDER OF THE GUNS

The outbreak of war in August 1914 came as something of an anticlimax to the vast fleets which both sides had created. Most prewar pundits had predicted a pitched battle in the North Sea, if not within hours of the declaration of war certainly within weeks. Others had predicted a sudden onslaught by German torpedo boats on the hapless British Home Fleet and some had even predicted a British preemptive strike on the German High Seas Fleet. All these views were wide of the mark and for some months virtually nothing happened.

The reason was simply that neither side intended to throw away its trump cards. The Germans lay secure behind thick minefields and the guns of the Heligoland fortress while the British battle squadrons seemed to have disappeared completely. In fact they had been sent to a prearranged rendezvous in the Orkneys, the huge natural harbor of Scapa Flow. As long ago as 1906 money had been voted to provide a new fleet base at Rosyth in the Firth of Forth which was better placed than Plymouth, Portsmouth or Chatham to dominate the North Sea. However by 1912 the fleet had grown to such a size that even Rosyth would be inadequate and it was decided to base the main fleet much further north, where it could react quickly to any attempt by the German fleet to break out into the Atlantic. Nothing was said publicly about this decision and nothing was done to prepare the lonely northern anchorage apart from providing a fuel depot, and right up to the last minute the Germans continued to imagine that the British were going to try a classic close blockade of their ports.

Moving the Home Fleet (given a more ancient and resounding name, the Grand Fleet, on the outbreak of war) from its peacetime southern bases to Scapa Flow was a mighty task of logistics and planning. Fortunately July 1914 had seen an experimental full mobilization of the fleet to test the arrangements for calling up reservists, and when the

First Lord of the Admiralty, Winston Churchill prudently cancelled leave because of the crisis following the Sarajevo murders, virtually the whole strength of the Royal Navy in home waters was ready and fully armed. In any case there were no plans for a secret German attack on the British fleet, and by the time the first U-Boats were gingerly pushing their way out of the Heligoland Bight the juicy targets they had expected were no longer around.

The Orkneys are treeless and forbidding but the Royal Navy did its best to create something resembling a fully equipped base as fast as it could. Light guns and searchlights were landed from ships and emplaced to cover entrances to the harbor and a boom of nets was provided. The island of Flotta was taken over for recreation, with soccer pitches, a nine-hole golf course and other amenities. To relieve the monotony each squadron of battleships was later sent to Invergordon for a month in rotation, where both repair facilities and amenities for the men were much better. On the spot repairs were done by labor accommodated in an elderly battleship, the *Victorious*, which had given up her 12-inch turrets to arm two monitors, and in 1915 her sister *Mars* took up the same duties at Invergordon.

Although the British showed none of the recklessness which German plans had counted on, the first use of their capital ships was a piece of opportunism which amply justified the risks taken. On 28 August 1914 a force of light cruisers and destroyers pushed into the Heligoland Bight to attack the German outposts, and ran into stiff opposition from light cruisers. Vice-Admiral Sir David Beatty was outside the Bight with four battlecruisers waiting to cover the withdrawal, but when Beatty heard that the British forces were likely to be overwhelmed he took the *Lion, Tiger, Queen Mary* and *Princess Royal* in at high speed, ignoring the threat from U-Boats and minefields. The arrival of these powerful ships brought the action to a dramatic end, as their

Previous page: The American predreadnought battleships *Ohio* (right) and *Missouri* pass the Miraflores Locks on the Panama Canal in 1915. The *Wisconsin* waits in the background.

Below: HMS *Tiger* was the best of the prewar British battlecruisers, but she still compared unfavorably with her German contemporaries. This photograph was taken toward the end of her career.

13.5-inch shells smashed into two of the light cruisers, the *Ariadne* and *Köln*, turning the tables on the Germans and extricating the British light cruisers and destroyers. Although the British staffwork had been abysmally poor the vigorous handling of the big ships by Beatty left the strategic initiative firmly in British hands, in the North Sea at least.

In the Mediterranean things went badly, when by a series of errors the German battle-cruiser *Goeben* escaped from two of the Mediterranean Fleet's battlecruisers, the *Indomitable* and *Indefatigable*. The *Goeben* and her escorting light cruiser *Breslau* eluded their pursuers and their arrival at the Dardanelles helped to force Turkey from her uneasy neutrality into the arms of the Central Powers. In theory the British ships should have been able to catch the *Goeben* quite easily for her leaky boiler tubes limited her speed to 22.5 knots, but the *Indomitable* was in very foul condition and the *Indefatigable*'s machinery was not in peak condition either. At times three of the *Goeben*'s boilers were out of action and

stokers were fainting from exhaustion in the bunkers but she finally reached safety on 10 August.

The British worried constantly about their margin of superiority over the High Seas Fleet, not realizing that the Germans were most reluctant to risk their ships in a headlong battle. As the British saw it the High Seas Fleet would sortie from its bases when it was at full strength, and would choose a moment when the Grand Fleet was at its weakest – a coincidence which was much harder to achieve than it sounded. As a result, when the dreadnought HMS *Audacious* struck a mine and sank off the coast of northern Ireland on 17 October 1914 the Admiralty went to extraordinary lengths to pretend that the ship was still in existence. Her name remained in the secret Pink Lists of dispositions and her pendant number continued to be listed in all signal books right through to November 1918 – implying that German spies might well have access to the very highest level of secret information. Unfortunately the ship had been seen sinking in broad daylight by the liner SS *Olympic* (with many Americans on board) and although the British Press accepted censorship under protest such antics did not inspire confidence in other Admiralty communiqués in the future.

Above: The imposing lines of HMS *Indefatigable* (1909–12) belied her flimsy protection, on the scale of an armored cruiser.

Top: The rusty and dirty *Seydlitz* arriving at Scapa Flow in November 1918. She bore little resemblance to the battlecruiser which had fought so proudly at the Dogger Bank and Jutland.

Left: Vice-Admiral Sir David Beatty, appointed to command the British battlecruisers in 1914. In no sense an intellectual, he nevertheless had sound tactical sense.

Comparison of Dreadnought Strength – August 1914

British		German	
4 *Iron Duke* Class		4 *König* Class †	
4 *King George V*		5 *Kaiser*	
4 *Orion*		4 *Helgoland*	
1 *Neptune*		4 *Nassau*	
2 *Colossus*		1 *Derfflinger* †	(battlecruiser)
3 *St Vincent*		1 *Seydlitz*	(battlecruiser)
3 *Bellerophon*		2 *Moltke*	(battlecruiser)
1 *Dreadnought*		1 *Von der Tann*	(battlecruiser)
1 *Tiger*	(battlecruiser)		
1 *Queen Mary*	(battlecruiser)		
2 *Lion*	(battlecruiser)		
2 *Indefatigable**	(battlecruiser)		
3 *Invincible*	(battlecruiser)		

* Plus HMAS *Australia*, available for service with the Grand Fleet
† *Derfflinger* and two *Königs* still completing

Above: The old German coast-defense ship *Beowulf* at sea in the Baltic in 1915. On both sides of the North Sea veterans performed humdrum duties but they used too much manpower.

Below: The battlecruiser *Lützow*, which served with the First Scouting Group from August 1915. Trouble with her engines meant that she did not become fully operational until March 1916.

Above HMS *Inflexible* (1905–08), one of the two victors in the Falklands battle. Although regarded as a capital ship, her scale of armoring was no better than the original armored cruisers.

Below right: HMS *Canada* had been ordered as the Chilean *Almirante Latorre* but on completion in September 1915 she was taken over for the Grand Fleet.

The British had a superiority of 22 battleships and 10 battlecruisers to 17 German battleships and five battlecruisers, but their position in reality was much better and bound to improve even more in 1915. For a start the Admiralty had acted promptly to prevent two new Turkish dreadnoughts from sailing. One was something of a freak, the *Sultan Osman I*: originally ordered for Brazil as the *Rio de Janeiro* she had no fewer than seven twin 12-inch turrets and was the longest battleship yet built. The other, the *Reshadieh* was a smaller version of the current *Iron Duke* Class, with the same armament of 13.5-inch guns. Renamed *Agincourt* and *Erin* respectively they were both outclassed by a bigger ship building for Chile as the *Almirante Latorre*, armed with ten 14-inch guns. As HMS *Canada* she joined the Grand Fleet in 1915. The five *Queen Elizabeth* Class with eight 15-inch guns were well advanced, with *Queen Elizabeth* herself almost

ready and *Warspite* only two months behind her. Eight more 15-inch gunned ships had been laid down in 1913–14, known as the *Revenge* Class, but they were coal-fired 21-knotters as the cost of the oil-fired 25-knot *Queen Elizabeth*s was felt to be excessive.

The Germans, in contrast, found that the army laid prior claim to the few 12-inch and 15-inch guns available and in any case the shipyards were not able to do more than complete the second of the *Derfflinger* Class battlecruisers, the *Lützow*, and the first of the four *Bayern* Class 15-inch gunned battleships by the end of 1915. Whereas the British were able to cancel three of the *Revenge* Class and stop further work on capital ships without affecting their margin of superiority, the Germans optimistically ordered four *Mackensen* Class battlecruisers with eight 13.8-inch guns and hoped to follow them with three 15-inch gunned ships, the *Ersatz Yorck* (replacement for the *Yorck*) Class. Had they been built they would have been formidable, but like the French *Normandie* Class of 1913 they succumbed to wartime shortages, primarily a lack of steel but ultimately of skilled labor.

The sinking of Rear Admiral Cradock's two old armored cruisers off Coronel by a German squadron under Vice-Admiral von Spee on 1 November 1914 led to one of the swiftest counterstrokes in the history of naval warfare. Three days later Vice-Admiral Sturdee hoisted his flag in the battlecruiser *Invincible* and she and her sister *Inflexible* left Cromarty Firth to go south to Devonport to take on stores.

British prestige had been dealt a heavy blow by the defeat at Coronel and the Admiralty was determined to bring Spee to action as fast as possible. After only a week the two battle-cruisers left for South America, but to avoid being reported by neutral or unfriendly ship-ping they took a circuitous route and stopped to coal at the Abrolhos Rocks in the South Atlantic, arriving at Port Stanley in the Falk-land Islands on 7 December. There they found the two survivors of the Coronel battle, the old battleship *Canopus*, and a scratch force of cruisers under Rear Admiral Stoddart.

The two big ships immediately started the arduous task of coaling and next morning while they were still engaged in the task lookouts sighted two German cruisers approaching Port Stanley. The *Canopus* had been put aground on mudflats to provide a steady gun-platform, and very soon her salvos were booming out to deter the attackers from any attempt to pick off the British ships as they left the harbor. Once the battlecruisers worked their way clear of the harbor the fate of the German squadron was sealed. It was the job for which the *Invincible* Class had been designed, and as the armored cruisers *Scharn-horst* and *Gneisenau* had not been docked for some time they were bound to be overtaken sooner or later. Sturdee knew this and settled down to a long stern chase, the two battle-cruisers belching clouds of coal smoke as they worked up to full speed with the British cruisers desperately trying to keep up with them.

Shortly before 1300 hours the first shots were fired at 16,000 yards. Spee gallantly hauled his flagship and her consort around to give battle and at the same time cover the flight of his three light cruisers, but with four British cruisers coming up their freedom would be short-lived. The weakness of long-range gun-nery was again demonstrated as it took the two battlecruisers four hours and over 1000 12-inch shells (an average of 73 rounds per gun, nearly the entire contents of their shellrooms) to sink the *Scharnhorst* and *Gneisenau*. Neither *In-vincible* nor *Inflexible* had yet received director gear but the real difficulty was trying to spot fall of shot at maximum range from a ship vibrating excessively at maximum speed. Once the range came down the shooting steadied, and even though both ships were hit by German shells of all calibers they suffered only slight damage.

Fisher had been recalled to the Admiralty as First Sea Lord in October 1914 at the insti-gation of Winston Churchill, and it would have been less than human for him not to claim that 'his' battlecruisers had saved the day at the Falklands. On the strength of this welcome victory he was able to overturn a Cabinet ruling that no new capital ships were to be laid down, and persuaded the Government that he could use material already assembled for the three cancelled *Revenge* Class battleships. Mindful of his coup with the *Dreadnought* he promised that the two ships, retaining their projected names *Renown* and *Repulse*, would be ready in only 15 months. This proved rather opti-mistic, but by judicious cutting of corners they were ready within 20 months – a staggering pace for such large ships of totally novel design.

The *Renown* and *Repulse* were built around available 15-inch turrets, and so displaced 26,500 tons and had only six guns. True to Fisher's ideals they were fast, just under 32 knots, but were ludicrously underprotected, with the same 6-inch belt given to the *Invincible* nearly a decade earlier. The Engineer-in-Chief wanted to try small-tube boilers and lighter machinery but to speed up design and building time the machinery of the last battlecruiser, HMS *Tiger* was duplicated with additional boilers. Whatever their faults they were handsome ships which reflected as much credit on British naval shipbuilding as the *Dreadnought* had.

Although given grudging permission to build two battlecruisers Fisher pressed ahead with three more ships which reflected his prejudices to an even more marked degree. To camouflage their identity they were described as 'large light cruisers' displacing over 19,000 tons but Fisher intended them to act as ultra-fast light battlecruisers to support his cherished scheme for amphibious landings in the Baltic. The first two were armed with two 15-inch turrets each and were capable of 32 knots but for the third ship the Elswick Ordnance Company was asked to design an 18-inch 40-caliber gun. This was a massive gun weighing 150 tons and firing a 3320-pound shell nearly 30 miles, and only single mountings could be produced if the dimensions of the twin 15-inch were not to be exceeded by an enormous margin. In fact the turret-ring was given the same diameter so that if the 18-inch gun proved a failure the ship could be rearmed later with twin 15-inch, but it proved to be just as successful apart from the colossal muzzle blast. The *Glorious* and *Courageous*, and particularly the *Furious* with her two single 18-inch turrets were probably Fisher's ideal warships, but with only 3-inch side armor they could never justify the term capital ship and it is hard to see what role they were expected to perform. The voluminous Fisher Papers give no hint; perhaps they were to lure German ships away from the beachhead but if not they

would be too clumsy for shore bombardment and given the crude fire control of the day two 18-inch or four 15-inch guns could not hope to hold the range in a fast-moving action.

Meanwhile the Grand Fleet was learning to live with the soul-destroying routine of life at Scapa Flow. In October 1914 there was a submarine scare in the Flow itself when one of the patrolling destroyers thought she sighted a periscope. Immediately the Commander in Chief, Admiral Sir John Jellicoe, ordered the Grand Fleet to be dispersed to a series of temporary bases in northern Ireland and the west coast of Scotland until permanent net defenses and minefields could be provided. For an anxious month the exit to the Atlantic was unguarded but the High Seas Fleet was not inclined to take advantage. Instead Admiral Hipper was instructed to lead his battle-cruisers on a raid across the North Sea to bombard Yarmouth in November 1914, fol-

lowed by a similar bombardment of the Yorkshire coast the following month. The idea behind these pinprick raids was to force the British to divide their fleet and so permit the High Seas Fleet to pounce on a weaker portion of it and fight at favorable odds. It worked, but only partially, to the extent that public outcry forced the Admiralty to move Beatty's battle-cruisers from Scapa Flow to Rosyth, where they were better placed to intercept such raids.

Intelligence of a similar raid led the Admiralty to order some of the Grand Fleet battleships and the Rosyth battlecruisers to meet off the Dogger Bank. This time a light cruiser sighted Hipper's battlecruisers and the result, on 24 January 1915, was the brief skirmish known as the Battle of the Dogger Bank. In a hectic stern-chase the flagship *Lion*, the *Tiger*, *Princess Royal*, *New Zealand* and *Indomitable* slowly overhauled Hipper's flag-ship *Seydlitz*, the *Moltke* and *Derfflinger* and

Below: The cage mast of the USS *Michigan*, damaged in a storm in January 1918. Most US dreadnoughts had such masts removed between the world wars and British-style tripods put in their places. As well as being stronger the tripods were more rigid so that there was less vibration interfering with delicate instruments.

Right: Rear Admiral Franz
von Hipper, commanded the
battlecruisers of the First
Scouting Group of the High
Seas Fleet at the Dogger
Bank and Jutland.

Below: The *Derfflinger*
(1912–14) taking on
ammunition.

last dynamo was short-circuited by water and
all light and power failed, bringing the ship to
a dead stop. Beatty transferred his flag as
quickly as he could but by the time he had
boarded the *Princess Royal* and resumed
control of the action Hipper's ships were
drawing out of range and it was hopeless to
continue the pursuit. The *Indomitable* took the
crippled *Lion* in tow and the whole force
returned to Rosyth. The *Blücher* had been
sunk but a magnificent opportunity had been
wasted.

Yet again firing at long range had proved
much harder than anyone thought, and the
Lion had expended 243 shells for only four
hits, two on the *Blücher* and two on the
Seydlitz. What nobody on the British side
knew was that the second hit had very nearly
sunk the *Seydlitz* when it plunged through the
quarterdeck and pierced the 9-inch armor of
the after 11-inch turret's barbette. Although
the shell did not penetrate, the explosion flung
pieces of armor through the ring bulkhead and
the flash ignited cordite charges in the gun-
house, the lower hoists and the handing room,
and even in the magazine itself. When the
crew of the handing room opened a door to
escape into the next turret the draught fanned
the fire into a blaze which swept through both
hoists. In all 62 full charges, some six tons of
cordite, were destroyed and gutted the after
part of the ship but fortunately fire parties
were able to flood the magazines before the
11-inch shells could be detonated by the
flames. The death roll came to 159 men and 33
more were wounded, and a court of enquiry
decided that the fire had been caused by too
many shells and charges being brought forward from the magazines – a common way of
speeding up the rate of fire.

Once both sides realized that there was
going to be no immediate titanic clash of fleets
in the North Sea they were able to relegate the
oldest battleships to second-line duties, for
their crews were needed to man the large
numbers of destroyers, patrol craft and minesweepers now necessary. The Germans used
their old coast-defense ships to protect the
Baltic coast against any Russian attack but by
the end of 1915 the manpower shortage was
acute and they began to be laid up. The British,
with many more commitments, kept most of
their older ships in commission, some of them
as flagships on foreign stations but others in
more exotic roles. Four of the old *Majestics*
were disarmed to provide turrets for a series of
monitors built in 1915 to bombard the German
army's right wing on the coast of Belgium, and
three of these disarmed battleships were then
used to transport troops to the Dardanelles.

the armored cruiser *Blücher*. The main weight
of British fire fell first on the *Blücher* bringing
up the rear and as a result of a later signalling
error the British ships turned aside to concentrate completely on the dying *Blücher*
although the German battlecruisers were being
caught.

The mix-up came at the moment when the
Lion was badly hit by several German shells. In
all she was hit 17 times but it was the last which
did the most damage. The shell, probably a
12-inch from the *Derfflinger*, which burst on
the 9-inch armor roughly on the waterline
abreast of the port engine room, blew in a plate
16 feet by 5 feet 9 inches and caused extensive
flooding. Although this flooding was brought
under control the ship listed about 10 degrees
to port and slowed down to 15 knots; finally the

Others served as guardships in east coast ports, like the German ships, while two were converted to repair ships for service at Scapa Flow and Invergordon.

It had been widely assumed before the First World War that the Royal Navy's main task would be to transport the small British army to fight in some distant theaters. 'The Army is the biggest shot which can be fired by the Navy,' expressed the idea in the simplest terms. This strategy had been quietly replaced by a 'Continental' strategy which required the British to fight alongside the French, but when this failed to produce the early decisive results hoped for British strategists began to think again of attacking the Central Powers where they were weakest. The obvious place to start was the Dardanelles, for it offered an alluring hope of knocking out the newest ally of Germany and Austria and at the same time opening a supply route to the Russians through the Black Sea.

The British and French Mediterranean Fleets had bombarded the outer forts of the Gallipoli peninsula on 3 November 1914 but the decision to mount a full-scale amphibious assault was not taken until the end of December. On 11 January 1915 Vice-Admiral Carden, the British flag officer in the Aegean, sent in his detailed plans for a three-stage assault; it envisaged a bombardment of the forts, sweeping the minefields in the Dardanelles and an advance by the combined fleets through the Sea of Marmora to the Bosphorus. Churchill, frustrated in his desire to bring the High Seas Fleet to action in the North Sea, gave the scheme his full support and even proposed to use some of the oldest battleships as expendable 'mine-bumpers.' Carden's requests included a large stock of ammunition for shore bombardment, 12 battleships, three battlecruisers and a mixed force of light cruisers, destroyers and sweepers.

So far so good, but at this point the planners allowed their enthusiasm to run away with them. The Admiralty War Staff Group even ordered the new battleship *Queen Elizabeth* to be sent out to calibrate her 15-inch guns against the forts! Not only was a new ship unlikely to make good target practice but the stocks of 15-inch shell were still too low, and it would have made much more sense to release one of the 13.5-inch gunned *Orion*s as there were ample reserves for that caliber. The final naval forces allocated looked formidable on paper:

Dardanelles Bombardment Force

British		French	
Queen Elizabeth	8 × 15-inch	*Suffren*	4 × 12-inch
Inflexible	8 × 12-inch	*Charlemagne*	4 × 12-inch
Agamemnon	4 × 12-inch 10 × 9.2-inch	*St Louis*	4 × 12-inch
Vengeance	4 × 12-inch	*Bouvet*	2 × 12-inch 2 × 10.8-inch
Albion	4 × 12-inch		
Cornwallis	4 × 12-inch		
Irresistible	4 × 12-inch		
Triumph	4 × 10-inch 14 × 7.5-inch		

Left: SMS *von der Tann* (1908–10), the first German battlecruiser, late in 1918 after removal of her torpedo nets.

In all they mounted some 170 modern guns from 5.5-inch upward, against 19 heavy guns in the outer forts, and among the latter the only modern weapons were two pairs of Krupp 9.4-inch guns at Cape Helles and Orkanie. It all looked too easy.

The bombardment began at 0950 hours on 19 February with an opening salvo from HMS *Cornwallis* at 12,000 yards, just outside the range of the most powerful Turkish guns. It proved much harder to hit land targets than anyone had thought, and at 1030 hours the ships were ordered to anchor in order to improve their shooting, but by 1400 hours the forts were still not silenced. When the *Suffren*, *Cornwallis* and *Vengeance* moved in to 5000 yards the smaller forts immediately opened fire on them and they were eventually forced to withdraw. It had not been realized before the war that it would prove so hard to spot the fall of shot on a low-lying coast devoid of prominent landmarks. Another miscalculation had been the effect of shellfire against coast-defense guns; contrary to expectation it was necessary to hit the gun itself or detonate the magazine. But above all the ships were restricted to a very low expenditure of ammunition; during the first day's bombardment 42 12-inch guns fired only 139 rounds, an average of 3.3 rounds per gun. By later standards this was laughable, and it stemmed from a fundamental ignorance of the problems of shore bombardment.

Six days of bad weather prevented the bombardment from continuing but on 25 February it started again. This time some battleships were allocated as flank markers, stationed on either flank to signal corrections for distance, and this improved the standard of shooting considerably. HMS *Queen Elizabeth*

distinguished herself by knocking out both the modern 9.4-inch guns at Cape Helles in only 18 rounds, while the older *Irresistible* took 35 rounds to knock out the Orkanie battery. That afternoon the trawler-minesweepers were brought forward to begin clearing the minefields and it was possible during the next few days to land parties of sailors and marines to destroy individual guns. However the inner forts proved a much harder nut to crack. The Straits become wider at this point, and the only deep water is well away from the shore so that the batteries were too far away for accurate spotting. The Allies had no modern aids to fire-control such as short-wave radio and the seaplanes used were so primitive as to be virtually useless.

In the first week of March three bombardments of the inner forts proved inconclusive. The *Queen Elizabeth* fired for four hours at 14,000 yards on 5 March at forts on the

Left: The old French battleship *Charlemagne* (1894–98) served with distinction at the Dardanelles and later in the Mediterranean.

Right: The *République* and her sister *Patrie* (1901–06) and the slightly later *Verité* Class formed the backbone of the French Fleet for many years.

Below right: The old French battleship *Gaulois* (1895–99) on 18 March 1915 sinking by the bows after being hit below the waterline by Turkish shells. She managed to limp as far as Malta for emergency repairs.

Below: The *Suffren* (1899–1903) survived heavy damage from Turkish gunfire at the Dardanelles in March 1915 but was torpedoed by a U-Boat in November 1916.

Above: Polishing the guns of a German battleship.

European side of the Straits, but as she fired an average of one round per gun per hour the results were negligible. Next day she fired against Chemenlik, a fort on the Asiatic side but the Turks had decided on a bold counter-stroke. Unnoticed they had moved the old battleship *Hairredin Barbarossa* (formerly the German *Kurfürst Friedrich Wilhelm*) down to the Narrows at Chanak, from where she could range on the *Queen Elizabeth*. Although her short 11-inch guns were quite outclassed by modern ordnance the mountings could elevate to 25 degrees, and with a steep angle of descent the shells could easily have penetrated the thin deck armor of the *Queen Elizabeth*.

The first three rounds were ignored by the *Queen Elizabeth* for it was thought that they had come from a mobile field-howitzer, but to be on the safe side she moved out 1000 yards. When the *Hairredin Barbarossa*'s shore spotting position was discovered the *Queen Elizabeth*, *Agamemnon* and *Ocean* fired at it until it was abandoned, but the observation party quickly established themselves in a new position. The bizarre duel started again and this time the Turkish ship found the range in only three rounds, hitting the *Queen Elizabeth* below the water-line three times. Fortunately for the British ship all three shells were stopped by the armor. It is quite conceivable that a lucky shot could have penetrated the decks and detonated a magazine, and realizing her danger the big ship moved out of range.

After the failure of the bombardments a final 'big push' was planned for 18 March, but this proved a disaster. No fewer than 18 French and British ships were assigned to bombard the forts in three successive lines:

Line A – *Queen Elizabeth, Agamemnon, In-flexible, Prince George, Triumph*
Line B – *Suffren, Bouvet, Charlemagne, Gaulois*
Reserve – *Vengeance, Irresistible, Swiftsure, Majestic*
Covering force for Minesweepers – *Cornwallis, Canopus*

In brilliant sunshine the two lines of ships opened a heavy fire and the Turkish batteries replied with equal resolution. Soon the fore-bridge of the *Inflexible* was set on fire, the *Gaulois* was holed badly below the water-line and the *Agamemnon* was hit a dozen times in less than half an hour. Just after 1400 hours the line of French ships turned starboard into Erenköy Bay on the Asiatic side, making way for the reserve line coming up behind. Little did anyone know that 10 nights earlier a small Turkish minelayer called the *Nousret* had slipped down the Straits to lay a field of 20 mines in the bay. Minesweepers had repeatedly swept there but finding only three mines had not realized that they were part of a line, and in any case it had been assumed that seaplanes would be able to spot minefields.

The *Suffren* turned to starboard successfully and made her way to the south but suddenly the second ship, the *Bouvet* lurched and a huge explosion shrouded her in steam and smoke. Then she rolled over and sank, taking 600 men with her. At first it was assumed that an unlucky shot from a heavy shell had set off her magazines, for she was an old and poorly protected ship, but two hours later the British battlecruiser *Inflexible* was rocked by another explosion. This time there was no doubt that it had been caused by a mine, and three minutes later the *Irresistible* reported that she had also been mined. The *Inflexible* was taking on water but managed to limp clear and made her way back to Malta for repairs. But when Commodore Keyes ordered the *Ocean* to try to take the stricken *Irresistible* in tow her captain refused to accept orders from a junior officer. Instead he started to steam back and forth, firing his guns aimlessly at the shore until he too felt an ominous explosion under his ship. With one ship sunk, two more sinking and a modern battlecruiser out of action it would have been suicidal to go on, and as Admiral de Robeck thought that the Turks had sent down drifting mines he ordered the ships to withdraw immediately.

The stalemate was now complete; the minefields could not be swept until the Turkish guns were silenced and the battleships could not silence the guns until the mines were swept. The fiasco of 18 March set in train the bloody debacle of Gallipoli, in which Australian, British, French and New Zealand troops spent their lives in a futile attempt to occupy both sides of the Straits. During the long agony of Gallipoli battleships tried vainly to provide covering fire for the troops ashore but they were too vulnerable. On the night of 15 May the *Goliath* was sunk by a Turkish torpedo boat off Morto Bay with the loss of 570 men, and then on 25 May the newly arrived *U-21* sank the *Triumph* off Gaba Tepe.

Left: Predreadnoughts of the US Navy's Atlantic Fleet (4th Division) at sea in 1917. Front to rear, *Louisiana, Kansas* and *New Hampshire*, seen from the maintop of the *Minnesota*.

Below left: The USS *Nebraska* in a unique 'dazzle scheme.' The idea of such camouflage was to mislead U-Boats about the speed and bearing of the target, not to make it invisible.

Below: The French pre-dreadnought *Saint Louis* took part in the first bombardment during the Dardanelles operation.

Right: A *Kaiser* Class
dreadnought steams across
the North Sea to surrender
after the armistice. These
24,400-tonners were armed
with ten 12-inch guns and
steamed at 20 knots.

Below: HMS *Lord Nelson*
was an 'intermediate
dreadnought' with a heavy
secondary armament of ten
9.2-inch guns in twin and
single turrets.

Two days later *U-21* arrived off Cape Helles
and spotted HMS *Majestic* with her torpedo
nets out and surrounded by colliers and
transports. In a superb demonstration of skill
Kapitänleutnant Hersing managed to slip two
torpedoes into the one-time pride of the
Victorian navy from a range of 400 yards. She
capsized slowly and fortunately loss of life was
slight but it was the end of battleship fire-
support for the ANZACs. Thereafter the
battleships remained at Mudros in case the
Goeben should try to break out, and fire-
support was left to destroyers until some of the
new monitors could be sent out from England.
After the withdrawal of Allied troops from
Cape Helles in 1916 the *Lord Nelson* and
Agamemnon took it in turns to maintain a watch
either at Mudros or Salonika against the
Goeben. As luck would have it both ships were
away when the German battlecruiser finally
sortied in January 1918.

The adventures of the *Goeben* merit a book

to themselves. She had propelled a reluctant
Turkey into war with Russia by bombarding
Sevastopol at the end of October 1914, despite
being nominally Turkish-manned as the
Yawuz Sultan e Selim. On 18 November she
fought a brief action against the Black Sea
Fleet, hitting the *Evstafi* four times and
receiving one hit in return before escaping in
fog. Then on 26 December 1914 she struck two
Russian mines off the Bosphorus and this
damage prevented her from playing a more
prominent part in fending off the Allied attacks
in March 1915. On 10 May she met the Black
Sea Fleet again, and this time the Russians did
much better. The *Evstafi* hit the *Goeben* three
times at the creditable range of 16,000–17,500
yards. On 8 January 1916 she met the new
dreadnought *Imperatritsa Ekaterina* and was
considerably discomfited to find that the
Russian 12-inch guns could elevate to 25
degrees and range out to 28,000 yards.
Although hit only by splinters it was an

unpleasant experience to be kept under fire at 25,000 yards, particularly as the German ship was slower than usual and her opponent was steaming very well.

Shortage of coal kept the *Goeben* in harbor from October 1916 onward, so that she could not take advantage of the declining morale of the Russian Fleet. Not until 20 January 1918 did she venture forth again, this time to strike at the British forces guarding the exit to the Dardanelles. At Imbros she disposed of the monitors *Raglan* and *M.28* with little difficulty but when she and the *Breslau* headed for Mudros both ships struck mines. For the *Goeben* it was the second mine, for she had already been slightly damaged by one just after leaving the Dardanelles, but then *Breslau* struck four more and sank. The *Goeben* tried to reenter the Straits but struck a third mine, and as she was now being bombed from the air by British aircraft in the confusion she ran aground off Nagara Point at 15 knots. There

she lay for six days while more bombing raids were made on her until finally the old battleship *Torgud Reis* and two tugs came down the Straits to get her afloat again. The bombs were too light to inflict any damage but the mine damage was not fully repaired until long after the war, and she was effectively put out of action until the Armistice.

When the Allies became involved in Greek affairs in 1916 the battleships were withdrawn to Salonika and subsequently four British pre-dreadnoughts were sent to bolster the Italian Fleet at Taranto. The Italians, uneasy at merely having parity with the Austro-Hungarian Fleet, demanded dreadnoughts but the Allied War Council treated this request with some acerbity. The Italians made good use of their ships for shore bombardment in the northern Adriatic and pushed ahead with the development of rudimentary motor-torpedo-boat tactics to bring the enemy to action. On the night of 9–10 December 1917 Luigi

Left: The old Austro–Hungarian coast-defense ship *Wien* (1893–96). She was torpedoed by an Italian MTB at Trieste in December 1917.

Left: The battered wreck of the Russian *Slava* lying in Moën Sound in the Gulf of Riga in October 1917 after she had been hit by the German *Kronprinz Wilhelm* and *König*.

Above: The first man to
sink a battleship by means
of a motor torpedo boat,
Commander Luigi Rizzo of
the Italian Navy.

Right: Rizzo's boat *MAS.15*
arriving at Ancona after
torpedoing the *Szent Istvan*,
11 June 1918.

Rizzo took *MAS.9* and his own *MAS.13* into Muggia Bay near Trieste to attack the old battleships *Wien* and *Budapest*. Two 18-inch torpedoes from *MAS.9* struck the *Wien* amidships and she sank quickly but the two fired by *MAS.13* missed the *Budapest* and hit a jetty.

Rizzo scored another outstanding success when in broad daylight on 10 June 1918 his *MAS.15* attacked the dreadnought *Szent Istvan*. The battleship was in company with her sisters about 10 miles west of Premuda Island when *MAS.15* and *MAS.21* approached unobserved and fired four torpedoes. The fate of the *Szent Istvan* is well known for as she rolled over she was filmed from another ship, and as the only World War I action movie of a ship sinking it features in most film libraries around the world. The Italian main fleet achieved very little but the efforts of Luigi Rizzo showed that given the right conditions and a vital element of daring the battleship was just as vulnerable as the torpedo-boat specialists had prophesied 30 years earlier.

Left: Some hours after *MAS.15* and hit her with a torpedo the *Szent Istvan*'s list became uncontrollable and she started to heel over.

Below: With dramatic suddenness the *Szent Istvan* rolled over on her beam end ends, taking 89 men with her.

6. THE ENIGMA OF JUTLAND

Jutland, or Skagerrak to the Germans, remains one of the most fascinating sea battles of all time. Even as early as 1930 Basil Liddell Hart complained that no battle had 'spilt so much ink,' and the rivers of ink have continued to flow since then. The reason is quite simply that so much was expected of it and yet so little resulted. The result was a baffling paradox: the Germans scored more material successes and cheated the British of a crushing victory, but did themselves no good whatsoever. What is much harder to comprehend is why it has taken historians so long to discover just what caused the various successes and failures. For Jutland was the first major clash of fleets in European waters since Lissa and the only full-scale battle between the two fleets in the entire war. It was also the largest sea battle to date, with 252 ships engaged, and the last one in which all the classic ship types played their parts, battleships, battlecruisers, armored cruisers, light cruisers and destroyers. Ironically it was also the first battle in which aerial reconnaissance played any sort of role.

The two sides had shadow-boxed in the North Sea since August 1914 largely because the German high command was only prepared to sanction hit-and-run raids on the east coast of England. But there was a faction inside the German navy which wanted a more positive policy and when in February 1916 Vice-Admiral Reinhard Scheer was appointed to command the High Seas Fleet he brought with him a new offensive spirit. This time he planned a coordinated trap for the Grand Fleet, sending U-Boats to mine the exit routes from Scapa Flow and Rosyth and then sending Hipper's First Scouting Group to trail its coat and lure Beatty's battlecruisers into the grip of the whole High Seas Fleet. It might have worked but for the fact that since the end of 1914 the Admiralty had been reading the majority of German cipher messages and when it was realized in London that the High Seas Fleet would be at sea Admiral Jellicoe was ordered to take the Grand Fleet out and rendezvous with Beatty's battlecruisers off the Skagerrak. The composition of the two forces was slightly different from usual because three of Beatty's older battlecruisers had been sent north to Scapa Flow for gunnery practice, and to strengthen the Battlecruiser Fleet it was given four of the *Queen Elizabeth* Class fast battleships.

Previous page: A German dreadnought firing a broadside.

Right: SMS *Von der Tann* at speed, some time in 1917. The combination of light and dark grey paintwork is a simple form of camouflage.

Everything still depended on chance, but when on 31 May 1916 two opposing groups of light cruisers turned to investigate a Danish steamer blowing off steam between them it was the prelude to a major action. The light cruisers were attached to their respective battlecruisers and each admiral closed the range eagerly in the belief that he could lure the enemy into the arms of his Commander in Chief. Fire was opened at over 24,000 yards, with six British battlecruisers against five German, for in his haste to get into action Beatty had left his 5th Battle Squadron of *Queen Elizabeth*s behind. They had deliberately been stationed 10 miles astern because Beatty did not want Hipper to refuse action in the face of greatly superior odds but it meant that these four powerful ships were late in coming into the fight.

The sea was calm but visibility was hazy when at 1546 hours the first salvos thundered out. Beatty had turned east to put himself between Hipper and the German bases while Hipper had turned southeast to draw Beatty back on the High Seas Fleet. Beatty's move was strategically sound but it put him at a disadvantage with the sun low in the sky behind his ships, silhouetting them clearly while leaving the German ships merging into the haze. To make matters worse dense clouds of coal smoke and cordite fumes began to roll across the water, adding to the difficulty of reading signals. By an error in signals from the flagship two British ships found themselves firing at the same target while the weakest ship, HMS *Indefatigable* was unsupported in a duel with the *Von der Tann*.

The action developed rapidly, with the *Moltke* scoring two hits on the *Tiger*, *Derfflinger* finding the range of the *Princess Royal* and *Lützow* hitting the *Lion*. At the rear of the line the *Von der Tann* and *Indefatigable*'s fierce duel reached a climax at about 1600 hours when the German ship scored three hits aft. The British battlecruiser lurched out of line with smoke pouring from her and sinking by the stern but before she could indicate the extent of her damage another shell landed near the forward 12-inch turret and a second hit the turret itself.

Below: HMS *Indefatigable* working up to full speed in the first phase of the Battle of Jutland, only minutes before she blew up with the loss of most of her crew.

Left: A hurried snapshot from HMS *New Zealand* shows the remains of the *Indefatigable* after she had blown up.

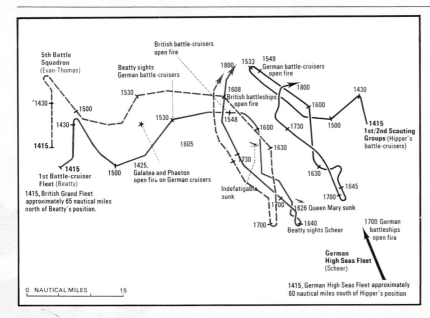

5th Battle
Squadron
(Evan-Thomas)

British battle-cruisers
open fire

1549
German battle-cruisers
open fire

1533

1800

Beatty sights
German battle-cruisers

1430

1800

1430

1600

1430

1500

1530

1600

1608
British battlesships
open fire

1730

1500

1415
1st/2nd Scouting
Groups (Hipper's
battle-cruisers)

1430

1548

1530

1605

1600

1415

1425,
Galatea and Phaeton
open fire on German cruisers

1630

1500

1st Battle-cruiser
Fleet (Beatty)

1730

1415, British Grand Fleet
approximately 65 nautical miles
north of Beatty's position.

1630

1645

Indefatigable
sunk

1700

1700

1626 Queen Mary sunk

1700 German
battleships
open fire

1700

1640

Beatty sights Scheer

German
High Seas Fleet
(Scheer)

0 NAUTICAL MILES   15

1415, German High Seas Fleet approximately
60 nautical miles south of Hipper's position.

Suddenly she blew up in a tremendous cloud of brown smoke and sheets of orange flame as her magazines detonated, and when the shower of debris cleared the *Indefatigable* had disappeared with nearly all hands. Her loss was caused primarily by the hits aft and the hits forward merely added to the catastrophe; at about 16,000 yards she was completely vulnerable to 11-inch shells. In only 20 minutes Hipper's ships had evened the odds.

Worse was to follow. At 1625 hours the *Derfflinger* shifted fire from the *Lion* to the *Queen Mary* and quickly straddled her. A 12-inch shell hit Q turret amidships and put the right-hand gun out of action, and about five minutes later another two shells hit, one near A and B turrets and the other on Q turret. Once again there was a huge explosion as the forepart of the *Queen Mary* vanished in a sheet of flame and smoke. Horrified onlookers saw the remains of the ship listing to port and sinking with the propellers still revolving, and then another·explosion obliterated her. Nine men survived out of 1285.

Some clue to the cause of the disaster was gleaned from the experience of the *Lion*. She was hit on Q turret just above the left-hand gun port by a shell which burst above the gun and killed or wounded all the gun crews. A fire broke out among the cordite charges near the guns but a fire party quickly ran a hose over the face of the turret and soused the wrecked turret with water. Yet a full 28 minutes after this the cordite in the turret burst into flame again and spread to the working chamber below. The combustion of the eight full cordite charges was so violent that Q magazine bulkheads were buckled and a venting plate blew out, allowing flames to enter the maga-

zine. Fortunately the magazine had already been flooded by a gallant Royal Marine turret officer, for if the magazine crew had merely closed the doors they would not have remained flash-tight. When the ship was examined later the burn marks in the hoist showed that the flame had jumped 60 feet, grim proof of how badly unstable cordite could behave. Many of the *Lion*'s 99 deaths occurred in this turret fire, which rivalled that in the *Seydlitz* at the Dogger Bank.

Help was at hand for the hard-pressed British battlecruisers, however, for the 5th Battle Squadron had caught up and was now able to open fire at extreme range. Within six minutes the flagship *Barham* was hitting the *Seydlitz* at nearly 19,000 yards and HMS *Valiant* was ranging on the *Moltke*, proof of how good their new 15-foot rangefinders were. The German ships could not reply at this colossal range, and the *Seydlitz*' report comments on how good the British fire-control had become. The only remedy was to make small shifts in course to throw off the British range takers but this had little effect. To make matters worse Beatty's destroyers launched a torpedo attack and hit the *Seydlitz*, tearing a hole 15 feet by 39 feet in her side. Yet such was the stout construction of German battle-cruisers that she was able to keep up full speed for a while.

The situation changed dramatically when Beatty learned from his light cruisers that battleships had been sighted, and two minutes later he saw for himself the mass of masts, funnels and smoke of the High Seas Fleet. This was just the way it was meant to happen, and he ordered his ships to turn about and go north, secure in the knowledge that the Grand Fleet was hurrying south at top speed to meet him. Yet again, however, the flagship's signal staff were too hasty and omitted to pass on the fresh instructions to the 5th Battle Squadron, leaving them to carry on firing at the German ships until they noticed Beatty's ships turning away. By the time they could be pulled around (1654) they were within range of the German battle line and suffered concentrated fire from Hipper's ships as well as the head of Scheer's line. The *Barham* and *Malaya* suffered several hits and took casualties but they and their sisters *Valiant* and *Warspite* fought back and avoided serious damage.

Relief for Beatty's hard-pressed ships was, however, almost at hand for Jellicoe had ordered the 3rd Battle Cruiser Squadron under Rear Admiral Hood to hurry ahead of the main fleet to reinforce Beatty. The three ships, *Invincible*, *Indomitable* and *Inflexible* came into action just as Beatty was starting to turn to the east across Hipper's bows in a deliberate attempt to prevent him from sighting the Grand Fleet too soon. Now the visibility favored the British for the first time and in

Left: An immense pall of smoke marks the grave of HMS *Queen Mary*, which blew up at about 1630 hours with the loss of 1276 men.

Right: Rear Admiral Hood, whose inspired handling of the 3rd Battlecruiser Squadron was tragically cut short when HMS *Invincible* blew up.

Below: The *Seydlitz* took an even worse battering than the one she had at the Dogger Bank, and had to be beached in the Jade River on her return.

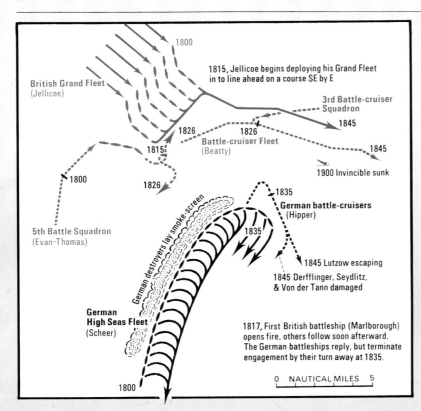

British Grand Fleet
(Jellicoe)

1800

1815, Jellicoe begins deploying his Grand Fleet in to line ahead on a course SE by E

3rd Battle-cruiser Squadron

1845

1826
1826
1845
Battle-cruiser Fleet (Beatty)

1815

1900 Invincible sunk

1800
1826

5th Battle Squadron (Evan-Thomas)

German destroyers lay smoke-screen

1835

German battle-cruisers (Hipper)

1835

1845 Lutzow escaping

1845 Derfflinger, Seydlitz, & Von der Tann damaged

German High Seas Fleet (Scheer)

1817, First British battleship (Marlborough) opens fire, others follow soon afterward. The German battleships reply, but terminate engagement by their turn away at 1835.

1800

0   NAUTICAL MILES   5

order to take the pressure off, Hipper ordered his destroyers to attack the British capital ships. But just as the light craft started to deploy at 1735 hours Hood's three battle-cruisers appeared out of the haze, 12-inch guns firing. Hood handled his big ships with great skill and within minutes they had inflicted crippling hits on the *Lützow*, and then reduced the light cruiser *Wiesbaden* to a wreck and damaged the *Pillau* seriously. The intervention of Hood's ships also distracted the German commanders by approaching from the east and accomplished their secondary mission of masking the approach of the Grand Fleet and giving it vital time for its deployment.

It must be remembered that it was now early evening and in the worsening visibility the rival commanders were working by guesswork and intuition. For Jellicoe in the Fleet Flagship *Iron Duke* the most urgent problem was to know just *where* the High Seas Fleet would be when it was sighted, and in which direction it

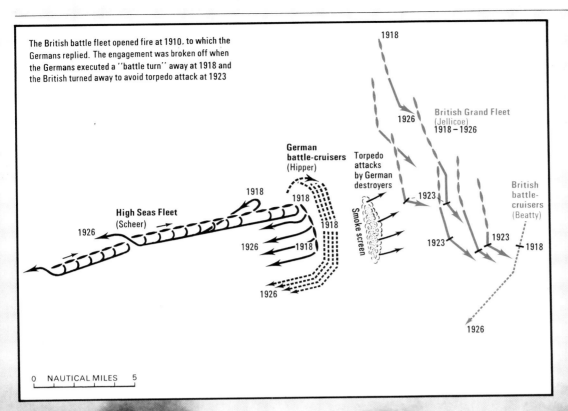

The British battle fleet opened fire at 1910, to which the Germans replied. The engagement was broken off when the Germans executed a "battle turn" away at 1918 and the British turned away to avoid torpedo attack at 1923

High Seas Fleet (Scheer)

German battle-cruisers (Hipper)

Torpedo attacks by German destroyers

Smoke screen

British Grand Fleet (Jellicoe) 1918–1926

British battle-cruisers (Beatty)

0 NAUTICAL MILES 5

Bottom left: HMS *Iron Duke* and an older dreadnought arriving at Scapa Flow in August 1914. Note the two-tone paint scheme, similar to that used by the Germans.

Bottom: The Grand Fleet deployment during the last daylight fighting of Jutland, with the *Royal Oak* and *Hercules* of the 6th Division, 1st Battle Squadron.

Below: This classic view of the Grand Fleet in cruising formation shows ships of the *Iron Duke* Class closest to the camera with other dreadnoughts in the background. From the modifications to the ships, it was taken in 1917 and it was in fact taken from the flagship *Queen Elizabeth*.

Left: The four *Queen
Elizabeth* Class battleships
of the 5th Battle Squadron
in action at Jutland, with the
flagship *Barham* leading.

would be steaming. From the flag-bridge he could see only seven miles, and if his own fleet was caught in the wrong formation or heading in the wrong direction his superiority in numbers and gunpower would be thrown away. The Grand Fleet was cruising in six parallel columns, a box formation to reduce the risk from submarine attack, and it had to deploy into line ahead to bring every available gun into action. Finally the fog of conflicting and garbled sighting reports resolved itself, the diminutive admiral studied the plot for no more than 10 seconds and then gave a series of orders for deployment on his port column. It was the right decision for at one stroke he interposed his fleet between the High Seas Fleet and its bases and brought his most powerful squadron into action first. Anything else would have involved immensely complex maneuvering or would have exposed the older dreadnoughts to heavy fire from the German battleships as they turned into line from the cruising formation.

As the majestic columns of ships wheeled in turn the two fleets finally sighted one another and ripples of orange flame flashed down the lines as the first ranging salvos were fired. Jellicoe had put the Grand Fleet in an ideal position, crossing Scheer's 'T' and concentrating fire on the head of the German line. The British battlecruisers and the *Queen Elizabeth*s also appeared out of the murk and took up their allotted position at the ends of the battle line but during this phase a sudden shift in visibility left the *Invincible* silhouetted against the setting sun, a perfect target for the *Derfflinger* and *Lützow*. At 10,500 yards she was an easy target and was hit five times. The fifth hit blew the roof off Q turret, and once again those ominous clouds of cordite smoke and coal dust billowed up as the *Invincible* broke in two. But Hood and the 1025 men who died had already made a most decisive contribution to the battle, and the ultimate loss of the *Lützow* later that night was directly attributable to HMS *Invincible*.

Below: The moment at which HMS *Invincible*'s 'Q' magazine detonated. Her forebridge and tripod mast are silhouetted by the flash of tons of cordite.

In direct contrast to the brilliant performance of Hood's 3rd Battle Cruiser Squadron the armored cruisers of the 1st Cruiser Squadron now immolated themselves in a totally pointless attack. Passing down the engaged side of the battle line and incidentally masking its fire with dense clouds of coal smoke, Sir Robert Arbuthnot led the *Defence* and three other cruisers in a headlong attack on the disabled light cruiser *Wiesbaden*. He paid for his stupidity with his life as the *Defence* blew up, while the *Warrior* was reduced to a sinking condition. All that had been achieved was a few hits on a ship which was already sinking. The *Warrior* and the other two were only saved from a similar fate by the arrival on the scene of the battleship *Warspite*, of the 5th Battle Squadron. She and her squadron were trying to follow Beatty's battlecruisers in taking the place at the head of the line but at a crucial moment her steering jammed, causing her to turn a complete circle near the *Warrior*. The German battleships at the head of the line

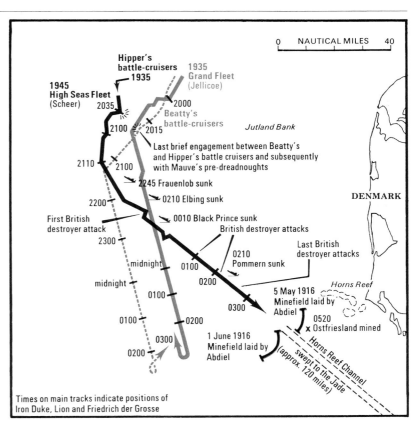

could not resist the chance of sinking a super-dreadnought and immediately switched fire from the *Warrior*. While the *Warspite*'s people struggled to cool down her overheated steering machinery she continued to circle, surrounded by shell splashes. The scene was nicknamed 'Windy Corner' by onlookers who were certain that she would be sunk but in fact she escaped serious damage and was able to get under control again. The distraction also gave the *Warrior* time to crawl away to safety, although she later sank. The *Warspite* had to leave the battle because she was losing boiler feedwater; she had been hit 13 times by the *Friedrich der Grosse* and the *Ostfriesland* and other ships.

The battle was now between the Grand Fleet and the High Seas Fleet, and the German gamble had failed. Jellicoe's insistence on constant gunnery practice showed in his flagship's shooting against the *König*. The German ship's war diary shows that the *Iron Duke* fired nine salvos in less than five minutes, of which seven shells hit. One Common Partially Capped (CPC) shell pierced the lower edge of the main belt armor and set fire to a number of 5.9-inch charges in a magazine. As with most German cordite it did not explode, in marked contrast to the behavior of British cordite. The loss of the *Indefatigable*, *Queen Mary* and *Invincible* might never have occurred if British propellant had been of the same standard.

The High Seas Fleet had only one course, its 'Battle Turn Away,' a difficult maneuver in which all ships turned 180 degrees simul-

Right: An artist's impression of the battlecruiser action at Jutland. The ship sinking is the *Queen Mary*.

Right: The main body of the Grand Fleet deploys into line from its cruising formation. Intensive practice was needed to ensure that such maneuvers could be performed accurately in the heat of battle.

taneously and steamed back on a reciprocal course. It achieved its aim of breaking contact but Jellicoe hauled the Grand Fleet around to the southeast at 1844 hours and again to the south 12 minutes later to keep it between the High Seas Fleet and its escape route. At 1908 hours Scheer blundered into the same trap, and this time his position was desperate, with the horizon apparently filled with hostile battleships. He ordered Hipper's battlecruisers on a 'death ride' supported only by the destroyers, to give the main fleet time to make another 'Battle Turn Away.' The destroyers were very roughly handled and the battle-cruisers took terrible punishment. The *Lützow* was hit five times and had two turrets put out of action. Five 15-inch hits on the *Derfflinger* disabled two turrets and started ammunition fires. A single 13.5-inch and two 15-inch hits on the battleship *Grosser Kurfürst* caused severe flooding. By now the German ships could hardly see the enemy, and the only hits scored were two by the *Seydlitz* on the *Colossus*, which caused minor damage.

The breakaway was successful, however, and the two fleets drew apart for the last time but there was to be one more engagement between the heavy ships. Beatty's battle-cruisers fired briefly at portions of the High Seas Fleet but at such long range as to be relatively ineffective. However one of HMS *New Zealand*'s shells put the after turret of the *Seydlitz* out of action – the first hit she had

Left: The somber face of Sir John Jellicoe shows the strain of commanding the Grand Fleet from 1914–16. His increasing pessimism led his replacement by Beatty, and he became First Sea Lord.

achieved out of the 420 shells she had fired during the action!

Up to this point Jellicoe could be assured of a handsome victory, for his fleet had proved superior in gunpower and tactics to Scheer's. The loss of life in the *Defence* and the three battlecruisers was tragic but the loss of the four ships had not affected the relative position. Twice the High Seas Fleet had been forced to run away from the Grand Fleet's guns and still the British were firmly in position across the route back to the German bases. Jellicoe's plan was to station his destroyers well astern of the battle fleet to prevent any attempt by Scheer to double back, and to maintain a cruising formation until daylight, when he

Below: The German *Helgoland* Class dreadnought SMS *Thüringen* fires on the British armored cruiser *Black Prince* (right) during the night action at Jutland.

hoped to continue the destruction of the High Seas Fleet. British tactics did not favor night action, for the very sound reason that it was a risky business, but they failed to take account of the absolute necessity for the Germans to fight at night in order to escape that annihilating battle in the morning. Furthermore the German navy had trained for night action whereas the British had not, with the result that when their ships met the British destroyers their reactions were quicker. What helped was that one of the British battlecruisers had imprudently passed the correct night challenge and reply to a sister ship by signal lamp when a German light cruiser was watching from the mist. The Germans had only part of the challenge but this was enough to make any British ships hold their fire for fear of hitting

Below: A 1913 view of the predreadnought *Preussen* leading the *Pommern, Hessen Lothringen, Hannover* and *Schlesien*, which formed part of the Second Squadron of the High Seas Fleet at Jutland.

one of their own ships.

In spite of these advantages the High Seas Fleet had a frightening ordeal before it finally smashed through the British light forces. When the predreadnought *Pommern* was hit by a single torpedo from HMS *Onslaught* her magazine exploded and she blew up with all hands. The light cruiser *Rostock* was torpedoed and the *Elbing* was rammed by the battleship *Posen* while twisting and turning to avoid torpedoes. The *Lützow* had strained her bulkheads by trying to keep up with the other battlecruisers, and when Scheer learned that she was drawing 70 feet of water forward and was unable to steam he ordered her to be torpedoed, and this was done at 0145 hours after destroyers had taken off her crew.

Once clear of the British flotillas the High Seas Fleet had a clear run home, apart from a mine which damaged the *Ostfriesland*. Although it had performed very creditably it had not won a victory of any sort, except in the sense of avoiding destruction. For the British it was a strategic setback: their fleet had been built at great expense to achieve an overwhelming victory and because that victory eluded them they had to continue with the ruinously expensive war on the Western Front. In relative strength they were hardly affected by the outcome; apart from the *Queen Mary* the losses had been in weak ships and these were replaced very quickly by much more powerful new construction. In the tactical sense they had won; Jellicoe was in possession of the battle area with his fleet intact while Scheer was hurrying back to base as fast as he could. The real blow for the British was to their pride, reputation and confidence. The press and public had been educated to believe that some sort of Trafalgar would happen, a battle of total annihilation, and when the Admiralty unwisely released the text of Scheer's report on the battle because Jellicoe's report was not complete, it was construed as proof of a

Above: Admiral Reinhard Scheer (nearest camera) and Prince Henry of Prussia watching exercises.

Below: HMS *Superb* opens fire on the High Seas Fleet at Jutland with the *Canada* astern. They formed part of the 3rd Division, 4th Battle Squadron.

British catastrophe. The losses were dramatic but against them could be set the *Lützow* and the *Pommern* and the light cruisers and the virtual sinking of the *Seydlitz* outside Wilhelmshaven. What was much more alarming was the weakness of training and command, for it soon became apparent that subordinates had not followed the correct procedures and had rarely shown any initiative in reporting sightings of enemy ships, assuming that Jellicoe already had the information. There had been plenty of gallantry but far too many mistakes and failures to interpret the spirit of the orders as well as the letter.

On the material side the worst shortcoming was undoubtedly the violent behavior of cordite propellant. None of the ships blown up could be described as well protected; three of them, *Invincible*, *Indefatigable* and *Defence*, were protected only on the scale of 1904 armored cruisers and the *Queen Mary* was built to meet an unrealistic Staff Requirement in which armor was deliberately sacrificed for speed and armament. The root cause of the explosions was that precautions for preventing flash from reaching magazines were nowhere near adequate to cope with the unforseen violence of the flash. Ironically the precautions

Left: The *Seydlitz* burns
furiously after being beached
outside Wilhelmshaven
because she could not enter
the harbor. She took four
months to repair.

Below left: Safely in dock at
last, the *Seydlitz* has no guns
in turret *Anton*.

Below: A vivid if inaccurate
artist's version of Jutland,
with the British 2nd Division
of the 2nd Battle Squadron
firing at the High Seas Fleet.

Above: Although only a 'throw off' gunnery exercise with practice ammunition, this gives an impression of what Jutland was like; a *Queen Elizabeth* battleship seen from the flagship, probably 1917.

in British ships were more elaborate than in German ships but because their cordite was more stable it never created an explosion.

A less obvious material failure was the weakness of the British armor-piercing shell, for this meant that the high-quality gunnery of the battleships of the Grand Fleet during the later phases of the action did much less damage to the German ships than it should have. In all there were 17 hits on German armor varying from 10–14 inches in thickness; of these one HE and three glancing Armor Piercing Capped (APC) shells had no chance of penetrating, but only one of the remainder penetrated the armor (*Derfflinger*'s barbette) and burst inside. Five more holed the armor without doing more than send fragments through and the other seven shells did not hole the armor at all. Against lighter armor British shells fared no better, and only one 15-inch APC shell penetrated the *Moltke*'s 8-inch upper belt armor at 18,000 yards. Some idea of the problems of long-range gunnery can be gauged from the expenditure of ammunition. An estimated 4480 rounds of all sizes were fired by the British capital ships and 3574 rounds by the Germans but they resulted in only 102 and 85 hits respectively. Nor were the results from torpedoes any more rewarding. The British fired 94 torpedoes, the Germans 105 and these achieved respectively six and two or three hits (excluding seven fired to scuttle ships).

The cordite and shell problems were tackled energetically by the Admiralty but inevitably it took time for new shells to be designed, tested and then manufactured in quantity.

Right up to the Armistice the *Revenge* and *Queen Elizabeth* Classes carried less than 30 percent of the new 15-inch APC shells in their shellrooms. There was no attempt to switch to solventless propellant like the Germans but quality-control during the manufacturing process was tightened and proper flash-tight scuttles were fitted to the ammunition hoists of all large ships. From April 1917 improved materials were used to make MC cordite and it was then possible to withdraw 6000 tons of defective cordite. The reequipment of the Grand Fleet was completed by March 1918.

Nothing could alter the inadequate scale of the battlecruiser's protection but what little could be done to reduce their vulnerability was done. They and the battleships underwent a series of modifications during the rest of 1916 and early 1917. This comprised principally the addition of 1-inch armor on the main deck amidships and around the barbettes, amounting to about 110 tons in the 12-inch and 130 tons in the 13.5-inch gunned ships. Needless to say, when the new *Renown* and *Repulse* joined the Grand Fleet after Jutland there was consternation at the long double row of scuttles, indicating just how little armor they had been given. Now that battlecruisers were no longer idolized for their speed Jellicoe ordered them back into dockyard hands for the addition of some 500 tons of extra armor. They had already received the standard 1-inch extra protection to the crowns of magazines while still completing, but this involved even more protection to go some way to remedying their deficiencies. Nevertheless they were treated with justified suspicion as 'tinclads' in the Grand Fleet, and because they had returned to the dockyards so quickly after commissioning they were dubbed HMS 'Refit' and HMS 'Repair.'

Both fleets completed their repairs as fast as they could but there was to be no second fleet action in the North Sea. The Grand Fleet continued its monotonous sweeps to no avail. On the one occasion that both fleets were at sea, in August 1916, two of the Grand Fleet's escorting light cruisers were torpedoed by U-Boats and as soon as Scheer learned that Jellicoe was at sea he returned to harbor. In November 1917 the British tried to repeat their August 1914 success with a raid into the Heligoland Bight. This time they sent a powerful force of light cruisers led by the *Courageous* and *Glorious*, with *Repulse* in support. The German dreadnoughts *Kaiser* and *Kaiserin* managed to get out of the Jade River to support their light cruisers and the British forces, already hampered by the extensive minefields, fought an inconclusive action. The absurdity of the design of the *Courageous* Class was demonstrated by their poor performance in action: between them they fired 149 15-inch shells and 180 4-inch but made only one hit on the *Pillau*. Similarly the *Repulse* fired 54 heavy shells for only one hit, but the worst performance was by the light cruisers, which fired the enormous total of 2519 shells and scored only three hits.

Given the problems of reequipping the Grand Fleet with new cordite and armor-piercing shells it was as well that 1917 saw no major action, but it did give Beatty (appointed to replace Jellicoe in 1917) the chance to improve standards of training. New techniques of concentrating fire were developed, ship-to-ship communications were improved and more searchlights aided night fighting. Range clocks were fitted on the masts to pass basic range estimates to ships ahead and astern, while deflection scales painted on the sides of turrets indicated the correct bearing, to avoid a repetition of Jutland when ships could not see the enemy simultaneously. The biggest change of all was the introduction of aircraft for spotting fall of shot. An attempt had been made at Jutland to use a seaplane from the carrier *Engadine* but this had failed because the seaplane's radio went out of action. From the autumn of 1917 capital ships were fitted with light wooden platforms on two gun turrets, one for a $1\frac{1}{2}$-Strutter reconnaissance aircraft and one for a Sopwith Pup fighter. Throughout 1915–16 the British attempts to intercept German forces had been frustrated by the excellent reconnaissance provided by Zeppelin rigid airships, and it was recognized that if the Zeppelins could be shot down before reporting the Grand Fleet's whereabouts the chances of another fleet action would be greatly improved.

Left: Two views of the superstructure of the Grand Fleet flagship HMS *Queen Elizabeth* late in 1917. Far left, the forward 15-inch turrets, bridge and foremast. Left, the searchlight platforms on the funnels, added after Jutland to improve night fighting.

Above: The *Florida* (left) and one of the *Wyoming* Class (right) of the US 6th Battle Squadron at sea in 1918, with the rest of the Grand Fleet in the background.

Below: Admiral Beatty and the crew of the *Queen Elizabeth* welcoming the American 6th Battle Squadron under Admiral Rodman at Scapa Flow, 1 December 1917.

In September 1917 the Americans sent a complete battle squadron across the Atlantic as part of their contribution to the Allied war effort. After a period in the south of Ireland they joined the Grand Fleet, and in a unique gesture of solidarity they were integrated into the Grand Fleet as the 6th Battle Squadron, adopting British procedures as far as possible. However the U-Boats' inroads into shipping were by now so serious that the US Navy was asked to send coal-burners rather than oil-

burners. The margin of superiority over the High Seas Fleet was now so large that the First Sea Lord offered to lay up a squadron of older dreadnoughts to ease the manpower shortage, but the Cabinet declined the offer for fear that it might offend the other Allies.

Frustrated and bored the men of the Grand Fleet might be but ultimately their morale survived intact whereas that of the High Seas Fleet crumbled. There were two reasons, first the tedium of inactivity in bases never

Scenes after the scuttling of the High Seas Fleet on 21 June 1919. Top to bottom: the upturned keel of the *Seydlitz*; crews of the scuttled ships waiting to be taken on board; fallen in on the quarterdeck; the prisoners exercising under guard.

designed to handle such large numbers of men and second, the steady drain of what might be called middle management, the younger officers and qualified petty officers and sailors, for service in U-Boats and torpedo boats. Their places were filled by conscripts and reservists but inevitably the gap between officers and men grew wider, and with so little time spent at sea there was plenty of opportunity for grievances to fester.

Nor could anyone be unaware of the worsening position of the civilian population as the Allied blockade steadily reduced the quality of foodstuffs. The pressure on the navy to do something to support their army comrades fighting in the east and in France was there all the time but as the Grand Fleet was more powerful than it had been before Jutland there was little that Scheer could do without risking a second and more disastrous encounter. The first rumblings of discontent were heard toward the end of 1917 and when at the end of the war Scheer and Hipper tried to take the Fleet to sea the sailors mutinied against what they believed to be a suicide mission.

When the end came it was almost un-expected. The German army requested an armistice and when the British insisted that the High Seas Fleet must be handed over this condition was met. On 21 November 1918 the might of the Grand Fleet mustered off Rosyth to see the unforgettable sight of 14 battleships and battlecruisers steaming in to give them-selves up. Between the long lines of ships they went, before being escorted to Scapa Flow for internment until the Peace Conference at Versailles should decide their fate. Beatty's signal was appropriately curt, 'The German flag will be hauled down at sunset and will not be hoisted again without permission.'

Once at Scapa Flow the German ships were cut off from the outside world, receiving only basic rations and letters from Germany. The British newspapers available gave very lurid accounts of the progress of negotiations at Versailles, and it was even suggested that the ships should be taken over by the Royal Navy and used to bombard the German coast if the German delegation refused to accept the terms of the peace treaty. Already humiliated

Below: The 26,000-ton hull of the *Derfflinger* heels slowly to port after being scuttled. She was not salved until 1936.

by their surrender and aware that the navy had apparently shown less concern for its professional honour than the army, the Imperial Navy officers decided on one last token gesture of defiance. Taking advantage of the temporary absence of the British battle squadron guarding them, the entire High Seas Fleet scuttled itself at its moorings on 21 June 1919. British boarding parties were able to save only the *Baden* and a few destroyers and the rest went to the bottom. It marked not only the end of a fleet but the end of the heyday of the battleship.

There would be bigger and better ships but never again would the battleship have the unchallenged prestige that she had enjoyed in August 1914. A breakdown of losses shows that out of 35 battleships sunk in 1914–18 only five were sunk by gunfire. The rest were torpedoed, 14, mined, nine, suffered internal explosions, six, or were simply scuttled, one. Less than a third of these were dreadnoughts, for despite the prestige of the big fleets it had been the old and vulnerable pre-dreadnoughts which had seen most service.

Table of Losses 1914–1918

Name/Nationality	Date	Cause
Austria–Hungary		
Wien	10 Dec 1917	Italian MAS-Boat, Trieste
Szent Istvan (D)	10 Jun 1918	Italian MAS-Boat, Premuda
Viribus Unitis (D)	10 Oct 1918	Italian limpet mine, Pola
France		
Bouvet	18 Mar 1915	Mined, Dardanelles
Suffren	26 Nov 1916	Torpedoed by U-Boat, Mediterranean
Gaulois	27 Dec 1916	Torpedoed by U-Boat, Mediterranean
Danton	19 Mar 1917	Torpedoed by U-Boat, Mediterranean
Germany		
Pommern	1 Jun 1916	Torpedoed by destroyer, Jutland
Lützow (D)	1 Jun 1916	Scuttled after gunfire damage, Jutland

Great Britain		
Audacious (D)	27 Oct 1914	Mined off Northern Ireland
Bulwark	26 Nov 1914	Blew up, Sheerness
Formidable	1 Jan 1915	Torpedoed by U-Boat, Channel
Irresistible	18 Mar 1915	Mined, Dardanelles
Ocean	18 Mar 1915	Mined, Dardanelles
Goliath	13 May 1915	Torpedoed by torpedo boat, Dardanelles
Triumph	25 May 1915	Torpedoed by U-Boat, Dardanelles
Majestic	27 May 1915	Torpedoed by U-Boat, Dardanelles
King Edward VII	6 Jan 1916	Mined off Cape Wrath
Russell	27 Apr 1916	Mined off Malta
Invincible (D)	31 May 1916	Gunfire, Jutland
Indefatigable (D)	31 May 1916	Gunfire, Jutland
Queen Mary (D)	31 May 1916	Gunfire, Jutland
Cornwallis	9 Jun 1916	Torpedoed by U-Boat, Mediterranean
Vanguard (D)	9 Jul 1917	Blew up, Scapa Flow
Britannia	9 Nov 1918	Torpedoed by U-Boat, Cape Trafalgar
Italy		
Benedetto Brin	28 Sep 1915	Blew up, Brindisi
Leonardo da Vinci (D)	2 Aug 1916	Blew up, Taranto
Regina Margherita	11 Dec 1916	Mined off Valona
Japan		
Tsukuba	14 Jan 1917	Blew up, Yokosuka
Russia		
Imperatritsa Maria (D)	20 Oct 1916	Blew up, Sevastopol
Peresviet	5 Jan 1917	Mined off Port Said
Slava	17 Oct 1917	Gunfire, Moon Sound
Svobodnaya Rossia (D)	18 Jun 1918	Scuttled, Black Sea
Turkey		
Messudieh	13 Dec 1914	Torpedoed by Brit submarine, Dardanelles
Hairredin Barbarossa	8 Aug 1915	Torpedoed by Brit submarine, Sea of Marmora

D:- Dreadnought

7. WASHINGTON'S CHERRY TREES

Such is the folly of mankind that the victorious Allies of 1918 were embroiled in a new naval arms race within a year. The balance of European power was a thing of the past, with the German navy eliminated and the United States emerging very rapidly as an industrial giant to whom all the other Allies except Japan were heavily in debt.

In 1916 Congress had authorized the construction of 10 battleships and six battlecruisers, bigger than anything previously built and armed with 16-inch guns. The purpose was quite simply to displace the British as the world's leading power, by building a navy 'second to none' which could, if necessary,

achieve its aims by force. The Japanese, aware that their island empire also stood in the way of such ambitions, and with plans of their own for the Far East, replied with the '8-8' Plan. This would provide eight new battleships and eight battlecruisers, also armed with 16-inch guns. This left the British completely outclassed, for although on paper they had an enormous superiority in numbers the older dreadnoughts were weakly armed and protected. Even the four fast battleships laid down in 1916 would be outgunned by the new Japanese and American ships as they only had 15-inch guns. From the Admiralty the projection for 1920 was distinctly bleak:

Previous page: The US Pacific Fleet on maneuvers. Front to rear, *New York, Texas* and *Wyoming*.

Admiralty Projections for 1920

Royal Navy		US Navy		Japanese Navy	
4 *Hood* Class	41,200 tons 8 × 15-in 31 knot	6 *S Dakota*	43,200 tons 12 × 16-in 23 knot	*Nos. 13–16*	47,500 tons 8 × 18-in 30 knot
2 *Repulse*	27,000 tons 6 × 15-in 31 knot	6 *Lexington*	43,500 tons 8 × 16-in 33 knot	4 *Kii*	41,400 tons 10 × 16-in 29.7 knot
5 *Revenge*	29,150 tons 8 × 15-in 22 knot	4 *Maryland*	32,000 tons 8 × 16-in 23 knot	4 *Amagi*	40,000 tons 10 × 16-in 30 knot
5 *Q Elizabeth*	27,500 tons 8 × 15-in 24 knot	2 *California*	32,000 tons 12 × 14-in 21 knot	2 *Kaga*	38,500 tons 10 × 16-in 26.5 knot
4 *Iron Duke*	25,000 tons 10 × 13.5-in 21 knot	3 *New Mexico*	32,000 tons 12 × 14-in 21 knot	2 *Nagato*	33,800 tons 8 × 16-in 25 knot
3 *K George V*	23,000 tons 10 × 13.5-in 21 knot	2 *Pennsylvania*	31,400 tons 12 × 14-in 21 knot	2 *Hyuga*	31,260 tons 12 × 14-in 23 knot
4 *Orion*	22,000 tons 10 × 13.5-in 21 knot	2 *Nevada*	27,500 tons 10 × 14-in 20.5 knot	2 *Fuso*	30,600 tons 12 × 14-in 23 knot
1 *Tiger*	28,500 tons 8 × 13.5-in 30 knot	2 *New York*	27,000 tons 10 × 14-in 21 knot	4 *Kongo*	27,500 tons 8 × 14-in 27.5 knot

Right: USS *Oklahoma* after her 1927–29 reconstruction with massive tripod masts. Later the funnel was raised to keep smoke away from the bridgework.

The Japanese had started work on the *Nagato* in August 1917 and her sister *Mutsu* in mid-1918, whereas the US Navy had been forced to defer its massive program. After laying down the *Maryland* in April 1917 nothing further could be done on her sisters because of the urgent need to provide destroyers to fight the U-Boats. They were finally laid down in 1919–20, followed by all six *South Dakota* Class in 1920 and the six *Lexington* Class battlecruisers in 1920–21. The Japanese similarly started the *Kaga* and *Tosa* and the battlecruisers *Amagi* and *Akagi* in 1920, with the *Atago* and *Takao* to follow in 1921.

The British *Hood* design was an expansion of the *Queen Elizabeth* design, a fast battleship rather than a battlecruiser, with 12-inch armor and a speed of 31 knots. Although she incorporated many new features such as inclined armor her design was too far advanced for all the lessons of Jutland to be incorporated, and as a result her deck protection was not adequate against long-range shellfire. The American and Japanese designs took even less account of battle experience, however. The *Nagato* and *Maryland* designs were fairly conventional in layout, with four twin 16-inch turrets, much like the British *Queen Elizabeth* Class. The *Kaga* and *Amagi* Classes had five twin turrets, two forward and three aft, with a raked flush deck, the *Amagi* having lighter

armor to give her higher speed. The last of the '8-8' program, numbered *13–16* and planned to be laid down in 1922, were to have eight 18-inch guns, making them the most powerful battleships in the world.

The projected American ships were equally unorthodox. The *South Dakota* Class had four triple 16-inch turrets, two forward and two aft, and the four boiler uptakes were in pairs athwartships but trunked together in one large and ugly funnel. The *Lexington* Class had boilers on two levels and at the initial design stage had seven funnels and ten 14-inch. Subsequently this layout was changed to eight 16-inch in twin turrets and five funnels, four of them side by side. In the final stage the thin multiple funnels gave way to two large, squat funnels, but the 7-inch belt remained inadequate for such large ships, even if it was inclined at 12 degrees to increase its resistance.

The British reply to this impressive series of ships was to give up all thought of completing the three modified sisters of the *Hood* – the Director of Naval Construction in any case regarded the design as inadequate in the light of wartime experience. The Admiralty hoped to order three new 43,500-ton battleships armed with nine 18-inch guns and a 48,000-ton battlecruiser with nine 16-inch in the 1921–22 financial year, and the same number of ships the following year. Probably because the existing British battlecruisers were so weakly

Above: The beautiful but ill-fated *Hood*, intended to be a 31-knot version of the *Queen Elizabeth*. Specially exempt from the tonnage limitation of the Washington Treaty, she remained the largest capital ship in the world until her loss.

protected the Board then revised the plan to four battlecruisers in the 1921–22 program and four battleships to follow in 1922–23. Other factors were the high speed planned for the American ships, 33.25 knots, and the fact that 16-inch guns could be built quickly by a number of armament firms, whereas 18-inch could only be handled by Elswick. There was wild talk of 20-inch guns or even 21-inch, but this stemmed from the head of Hadfields, the shell-makers, who boasted in 1920 that they were in the market for such projects.

The 16-inch gun finally chosen for the battlecruisers was a conventional wire-wound gun weighing 108 tons but in an excess of zeal to copy German ideas the previous sound combination of a heavy shell and low muzzle velocity was replaced by a light shell and high velocity. Trials showed that this produced loss of accuracy and stripping of the rifling, and muzzle velocity had to be reduced from 2700 feet per second to 2575 feet per second. The 18-inch Mk II was longer than the original Mk I in HMS *Furious*, mainly to improve its ballistics, and following the usual clumsy attempts at deception all correspondence referred to it as a 16-inch/50 caliber. This time there was to be a departure from the standard wire-bound construction and three prototypes ordered at the end of 1920 were to be wire-wound, half-wire and all-steel respectively. A fourth gun ordered slightly later was modelled

on Krupp principles to provide further information. Triple turrets had never been favored by the Royal Navy but weight considerations in the new designs made them imperative, and as British industry had provided most of the design expertise for the triple 12-inch turrets in the Italian and Russian dreadnoughts there were likely to be few technical problems. It was even suggested that a triple 12-inch should be taken out of the Soviet dreadnought *Volya* (ex-*Imperator Aleksandr III*) while she was under British control in the Crimea in 1920 but eventually a test-mounting with three 15-inch barrels was installed in a monitor for trials.

Other countries were working on super-heavy guns. The Japanese built two 19-inch to test but one burst during trials. The US Navy built one 178-ton 18-inch/48 caliber but like the Japanese guns it was only for evaluation. The French also started work on a 17.7-inch gun in 1920 but it never got beyond the design stage, but there could be little doubt that a new generation of battleships was in the offing, displacing around 50,000 tons and armed with 18-inch guns. It may have made sense to firebrands in Washington and Tokyo in 1916 but when the brief post-war boom collapsed and the inevitable slump had followed such megalomania began to look less and less attractive. The Americans were now in the worst position, having committed themselves

Below: Design sketch for 43,200-ton *South Dakota* Class, which were laid down in 1920–21 but cancelled in 1922. The peculiar funnel arrangement was dictated by the machinery layout.

to a program to build 1916-vintage designs in competition with the Japanese and the British. The British for once had the edge over the others because they alone had unfettered access to German designs for trials and a wealth of wartime experience. The Americans would be spending millions of dollars on ships which were already outclassed, a fact not likely to escape the attention of Congress. Indeed the mood on Capitol Hill was such that many senior naval officers doubted that, in the new mood of isolationism, any ships could be built at all.

There were other forces at work to stop the program. The Japanese, who had fought so hard to expel the Russians from their foothold in Manchuria in 1904–05 were not prepared to see a new Port Arthur built to threaten their sea power in the Far East. When the United States announced its intention to fortify Cavite in the Philippines Japan made it clear diplomatically but nonetheless firmly that this would lead to war. To complicate matters the British were bound by treaty to come to Japan's aid if she was attacked, so that the price of that navy second to none began to look even more expensive. There was only one way out, a negotiated mutual reduction of the naval programs, and President Harding instructed Secretary of State Charles Evan Hughes to convene an international naval disarmament conference.

The Washington Conference, as it is always known, opened on 12 November 1921 with the United States, Great Britain, Japan, France and Italy represented by powerful delegations. Germany was not invited, for her truncated coast-defense navy was governed by the Treaty of Versailles, and the anarchy prevailing in Russia meant there was little point in asking the Russians or Soviets. The basic proposals put forward by the United States were the scrapping of older capital ships without replacement, abandoning the enormous new ships under construction and placing limits on the size and numbers of new ships for a fixed period. Arising out of these limits on naval expansion there was also to be a fixed ratio of strength between the five major navies. They were revolutionary proposals for although arbitrary political limits had been placed on the size of individual ships before, this was the first attempt to regulate the growth of specific warship types and to dictate universal standards irrespective of each navy's strategic and tactical requirements.

The conference had far-reaching influence on all classes of ships but here we can confine ourselves to capital ships. The British accepted that there should be parity with the United States but the Japanese were furious when they were bracketed with the French and Italians at three-fifths of the British and American navies. Both the Japanese and the Americans wanted

above all to save what they could out of their huge programs; Japan wanted to avoid scrapping the nearly complete *Mutsu* while the United States wanted to keep the *Maryland* and her three sisters. As all three navies had or intended to build 16-inch gunned ships that was chosen as the maximum gun-caliber, but fixing the size of ships was harder. The Americans were happy to settle on 32,000 tons, the size of the *Maryland* Class, but the British, who had already instructed their designers to examine the possibilities of reducing the 48,000-ton battlecruisers felt that it would not be possible to build a balanced design (with adequate protection against 16-inch shells) on less than 35,000 tons.

There was also the knotty problem of how much new construction to scrap. The Japanese and Americans were in a stronger position than the British for they had a large number of hulls. The Royal Navy could only offer to scrap the *Hood*, which had just been completed, and accordingly much was made of the so-called *G.3* battlecruiser design, which if it were built would have a big margin of superiority over the latest American and Japanese ships. Finally the delegates agreed that Japan would keep the *Mutsu*, the United States would keep three of the *Maryland*s and the British would keep the *Hood* and be permitted to build two new ships to the 16-inch, 35,000-ton limit. In addition four large battlecruisers, two in the United States and two in Japan, were allowed to be completed as aircraft carriers, and the British were allowed to convert the smaller *Courageous* and *Glorious* as well. Finally the Treaty for the Limitation of Armament was signed in Washington on 13 December 1921. Although in no way part of it, the British had also been persuaded by strong American pressure not to renew the Anglo-Japanese Treaty.

The British were now able to press ahead with preparation of the two new battleships allowed by the Treaty. In 1921 they had conducted exhaustive trials on the *Baden*, the most powerful German capital ship completed,

which had been beached successfully at Scapa Flow after the scuttling attempt. Using the monitors *Erebus* and *Terror* to fire deliberately aimed and spaced shots at various angles and ranges the DNC's department examined the *Baden* at great length. These trials confirmed that the latest APC shells were fully efficient and it was also possible to test aerial bombs by placing them at various points on the decks. Further trials were carried out on the 'Chatham Float,' an experimental caisson on which sections of the *G.3* design's armor and structure were built.

The ships which resulted were remarkable by any criterion. The armament and scale of armoring of the 48,000-ton *G.3* design was retained at the expense of speed. By reducing the power by 75 percent and by reducing the length of the armored citadel it was possible to 'lose' 14,000 tons, but at the cost of grouping all three triple 16-inch turrets forward, B superfiring over A and X immediately abaft the barbette of B. This implied a risk of a lucky hit putting all three turrets out of action but as the scale of protection was heavy the risk was small. The armor belt was sloped and mounted internally to improve underwater protection and a novel form of 'water protection' or liquid-loaded layer was adopted. This had first been used in British monitors in 1915 and

Below: It is easy to see why *Nelson* and her sister *Rodney* were likened to oil tankers but they did not differ radically from the original 48,000-ton *G.3* design.

later in capital ships but purely as an external appendage to the hull. Incorporating it into the hull provided the same degree of protection but without affecting speed.

The new ships were named *Nelson* and *Rodney*, and when they appeared in 1927 their bizarre appearance caused much adverse comment. To the newspapers they were facetiously known as the 'Cherrytree Class' (because they were cut down by Washington) while matelots called them 'Nelsol' and 'Rodnol' because the stubby funnel aft and superstructure and turrets forward reminded them of a class of naval oilers with names ending in 'ol.' But they were soundly conceived ships, reflecting all the hard-won experience of the recent war. Good freeboard made them seaworthy in all weathers while for the first time

the secondary 6-inch were in twin turrets to protect them from spray and blast. There had been many complaints from bridge personnel about draughty bridge platforms and so *Nelson* and *Rodney* were given a massive tower structure enclosing all the various signalling and lookout positions, admiral's bridge and steering position. Other navies pointed out that 'Queen Anne's Mansions' offered a conspicuous target but the Admiralty rightly felt that spacious and secure bridge accommodation made for greatly increased efficiency and also provided a more rigid platform for fire control instruments.

At the Washington Conference there had been much wrangling over a new definition of displacement. Previously navies had measured warships in 'normal' or 'full load' tonnage,

Below: HMS *Rodney*'s massive freeboard was partly caused by the extra buoyancy when her 'water protection' was empty in peacetime.

normal being the average operational condition and full or deep load including the maximum load of fuel, feedwater, stores, ammunition and crew that the ship was capable of carrying. The British for example used a 'legend' fuel load, a quarter or third of the maximum, and other navies varied from one-third to two-thirds fuel and added in reserve feedwater for the boilers. At Washington the British maintained that fuel should be excluded from the basic tonnage, because their ships required a heavier fuel load to operate world-wide. Although this contention had more to do with the British desire to conceal the 2000 tons of water which would fill the *Nelson* and *Rodney*'s underwater protection scheme, it did lead to the new definition of standard displacement, which included ammunition and stores but not fuel and feedwater.

The battleship was not only threatened by disarmament treaties. The new devotees of air power claimed that she could be sunk with ease by bombing, and the more extreme exponents of the doctrine even went so far as to demand the abolition of navies. In a time of shrinking post-war budgets the rival claimants became even more strident and none more so than Brigadier General William Mitchell, then an influential but subordinate officer in the US Army Air Service. By 1920 his campaign for an independent air force on the lines of Britain's new Royal Air Force was reaching a crescendo of invective against the 'battleship admirals,' and he demanded a full-scale test of bombers versus surrendered German battleships. Tests made by the US Navy against the old pre-dreadnought *Indiana* in 1920, he maintained, had been kept secret to conceal the effect of bombing.

After a massive publicity campaign which included leaking photographs of the *Indiana* trials to the foreign press Mitchell was allowed to conduct a bombing exercise against ex-German ships in June–July 1921. In the first tests the AAS pilots had no trouble finding the radio-controlled *Iowa* and on 22 June they sank the submarine *U.117* with ease. The destroyer *G.102* went down as rapidly on 13 July and even if the light cruiser *Frankfurt* stood up better to 600-pound bombs she still sank convincingly. But the real prey of the airmen was the dreadnought *Ostfriesland* for Mitchell knew that nothing but a battleship would provide the proof he was after. The test was scheduled for 20 July and Mitchell took care to put about stories that the somewhat battered and neglected dreadnought had been known as the 'Unsinkable *Ostfriesland*' in the German navy. In fact an inspection four days earlier had found all watertight doors and hatches open, and she was in poor shape otherwise.

The first attacks were inconclusive because only nine of the 33 230-pound bombs dropped hit, and of those only two burst. At 1542 hours the pride of the Army Air Service, the Martin twin-engined bombers, appeared carrying two 600-pound bombs apiece, followed by a group of navy seaplanes carrying two 550-pound bombs. After five hits had been scored (two by army planes and three by the navy out of a total of 19 bombs dropped) a party of observers went aboard the *Ostfriesland* to inspect the damage. They found some flooding but not on a scale which would not have been checked by damage control parties if the ship had been manned. So insignificant was the damage that there was even talk of cancelling the remainder of the tests and allowing the *Pennsylvania* to sink the *Ostfriesland* by gunfire. In desperation Mitchell ordered his airmen to do whatever was needed, just as long as they sank the ship next day.

The US Navy was particularly anxious that this expensive full-scale test should not be wasted on a public relations stunt for they genuinely wished to learn what sort of damage heavy bombs could do to deck armor and structure. It had therefore been planned that as soon as a hit was observed on the *Ostfriesland* a halt would be called to allow observers to inspect the damage, as before. But this time, when the first army bomber scored a hit with a 1000-pound bomb a second bomber arrived and dropped another bomb nearby. Even when Admiral Wilson personally radioed the army pilot to stop his bombing run he was coolly told that he, the army pilot, would advise the admiral when the observers were free to go on board. Four more bombs followed, scoring two more hits before an apoplectic Wilson was able to get the army pilots to obey orders.

It was beginning to look as if Mitchell would not get his spectacular kill in front of the cameras, for the rules of the test allowed the army to drop only three of the largest bombs, 1-tonners. By now desperate, Mitchell ordered his bombers to ignore the rules and keep on attacking until they got two hits, and told Wilson of his intention just before the bombers went in. As his aviators had orders not to hit but to get near misses this was simply a lie to conceal his intentions. Even with such flagrant insubordination it still might have ended in an humiliating fiasco for Billy Mitchell. The first three bombs missed. The fourth appears to have glanced off the forecastle but the *Ostfriesland* was bracketed by the fifth and sixth, and these apparently opened her seams. She was now listing by the stern and gradually sank lower in the water. At 1241 hours, just 23 minutes after the first bomb had been dropped,

Left: The 'Mitchell Tests' against the *Ostfriesland*, 21 July 1921. An Army Air Service bomb explodes alongside the ship.

Below left: The stricken *Ostfriesland* begins to sink.

Bottom: The ram bow of the *Ostfriesland*, the last part of her remaining above the surface.

the by now very thoroughly battered *Ostfriesland* gave up the struggle, rolled over to port and sank without trace.

It would have been surprising if such a charade had not caused bitter inter-service rivalry. Mitchell kept asking for his court-martial and finally got it, but what exactly was proved by his performance off Hampton Roads on 21 July 1921 is debatable. Subsequent tests by the RAF against the old radio-controlled target ship *Agamemnon* were not so conclusive. Later war experience was to show that Mitchell's bombers were attacking at a ridicu-

lously low level, and his claim that it was easier to hit a moving target than a stationary one can be dismissed as twaddle. Nor could it be said that the ship was in a proper watertight condition; her flooded boiler-room and condenser compartment from the damage of the previous day had reduced her margin of stability and it needed only the blowing of her unprotected glass scuttles on the lower deck to start progressive flooding. In the words of a modern authority, 'Mitchell could have sunk the *Ostfriesland* with a carpenter's hammer.'

The 'bomber versus battleship' controversy was a sterile debate which did little credit to anyone. In Britain it took the form of cost comparisons, with Lord Trenchard and his supporters claiming that 1000 bombers could be built for the price of one battleship. So frequently was this piece of arithmetic used that the Admiralty was finally forced to produce figures, but unfortunately these showed that only 37 bombers could be built for the money. Worse, the bitterness of the quarrel blinded people to the fact that dive-bombing was a much more dangerous method of attacking battleships than high-level bombing. Although the Japanese and Americans both realised this and turned to dive-bombing in the 1930s the RAF ignored dive-bombing and became obsessed with high-level bombing, an aberration which led up the blind alley of strategic bombing. Had air power been seen as an ally of sea power rather than an alternative talented men like Billy Mitchell might have served their country even better than they did.

With a 10-year holiday or moratorium on battleship building the leading navies turned their minds to modernizing the capital ships left to them under the treaty. The Americans were quick off the mark for they were dissatisfied with their early dreadnoughts. Starting with the *Utah* and *Florida* nearly $8,000,000 was spent on fitting anti-torpedo bulges, giving them new machinery and converting the boilers to burn oil. The *Arkansas*, *Wyoming*, *New York* and *Texas* were similarly treated and like the earlier pair became single-funnelled ships in the process. From the *Nevada* Class onward modernization was much more comprehensive because of the ships' greater fighting value. They were given massive tripod masts in place of the traditional basket or cage masts which had been standard since 1911. The three *New Mexico* Class were not taken in hand until the 1930s and so received a later type of modernization which gave them a different appearance, with one tall funnel and much bigger bridgework than any previous battleship. The *California* and *Tennessee* and the *Maryland* Class were the only

Right: In 1929–32 the Japanese battlecruiser *Hiei* was disarmed to comply with the Washington Treaty, losing one 14-inch turret and five boilers.

Right: The *Fuso* was partially modernized in 1927–28 with extra bridge-work and a tall cowl on the forefunnel.

Right: The *Ise* was similarly modernized in 1929–30, and subsequently both classes underwent total recon-reconstruction.

ones not to receive major peacetime recon-structions, being newer ships.

The Japanese were equally determined to use the reconstruction clauses in the Washington Treaty to maximum advantage and started on the *Kongo* Class battlecruisers. Not only were they given oil-fired boilers but also anti-torpedo bulges and higher elevation for the 14-inch guns. In the mid-1930s they underwent an even more comprehensive modernization, turning them into fast battleships capable of escorting aircraft carriers. The speed was raised to 30 knots by lengthening the hull and installing more powerful machinery and a formidable antiaircraft armament. Like the American ships they were given catapults for launching reconnaissance floatplanes, vital for the vast distances of the Pacific before the advent of radar.

The *Fuso* and *Hyuga* Classes were converted to oil fuel and received a massive reconstruction in the mid-1930s, followed by the two *Nagato* Class. A feature of all these Japanese reconstructions was the massive 'pagoda' bridge structure, in reality a series of platforms around a hectapod mast.

Operational Capital Ships Post–1921

France

Diderot, Condorcet, Vergniaud

Jean Bart, Courbet, Paris, France

Bretagne, Provence, Lorraine

Great Britain

Hood

Renown, Repulse

Royal Sovereign, Royal Oak, Revenge, Ramillies, Resolution,

Queen Elizabeth, Warspite, Barham, Malaya, Valiant,

Tiger

Iron Duke, Marlborough, Emperor of India, Benbow

King George V, Ajax, Centurion, Thunderer [1]

Italy

Napoli, Regina Elena, Roma, Vittorio Emmanuele

Dante Alighieri

Conte di Cavour, Giulio Cesare [2]

Andrea Doria, Duilio

Japan

Kongo, Kirishima, Hiei, Haruna

Fuso, Yamashiro

Hyuga, Ise

Nagato, Mutsu

Soviet Union

Oktyabrskya Revolutsia, Parizhkaya Kommuna, Marat [3]

Turkey

Yawuz Sultan e Selim

United States

Florida, Utah

Arkansas, Wyoming

New York, Texas

Nevada, Oklahoma

Pennsylvania, Arizona

New Mexico, Mississippi, Idaho

Tennessee, California

Maryland, Colorado, West Virginia

[1] All to be scrapped on completion of *Nelson* and *Rodney*
[2] Refitting of salved *Leonardo da Vinci* cancelled 1923
[3] Ex-*Gangut, Petropavlovsk* and *Sevastopol*; *Poltava* damaged beyond repair and laid up.

The Washington Treaty had taken special account of the problems faced by the Italians and French. The Italian dreadnoughts were lightly armed and even the later French dreadnoughts were poorly armored. Italy had been forced to delay four 15-inch gunned fast battleships due to shortage of materials during the war and had cancelled three but even the surviving hull had to be sold in 1920 during a financial crisis. France had stopped all work on the *Normandie* Class in August 1914 and by 1920 it was obvious that the design was obsolescent, although the hull of the *Béarn* was kept for conversion to an aircraft carrier. To compensate for the lack of modern battleships Italy was to be allowed to build 70,000 tons of new ships and after 1927 France could lay down two new capital ships.

The Italians, true to their tradition of innovative design, achieved the most remarkable transformation of their old battleships. The *Conte di Cavour* and *Giulio Cesare* were gutted to allow the installed power to be nearly trebled and a completely new superstructure was built. The midships 12-inch turret was removed and to compensate for the loss of firepower the guns were bored out to 12.6-inch caliber and given more elevation. Additional deck armor was provided as was a novel system of underwater protection. Known as the Pugliese system from its inventor, it relied on a liquid-filled cylinder inside an anti-torpedo bulge, on the principle that a torpedo explosion would expend all its force in crushing the cylinder. In practice it was no more successful than other systems because it was impossible to keep the cylinder watertight after an explosion. The *Duilio* and *Andrea Doria* were given a very similar reconstruction in 1937–40 which differed only in the provision of a better anti-

Below: Although the US Navy had pioneered the catapult, for some time after the war turret platforms continued to be used to launch aircraft such as this Nieuport 28.

Right: The French *Lorraine* (1912–16) survived World War I and was modernized and reconstructed in the 30s. She was not scrapped until 1953.

Right: The 'pocket battleship' or *panzerschiff Admiral Graf Spee*. She carried six 11-inch guns on a displacement of 11,700 tons but was protected by only 3-inch side armor and steamed at 26 knots.

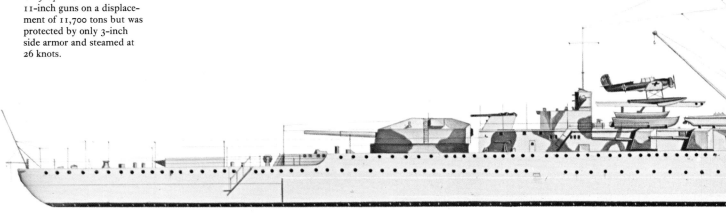

Right: The Italian *Littorio* Class were among the fastest battleships in World War II but they flagrantly exceeded the 35,000-ton limit.

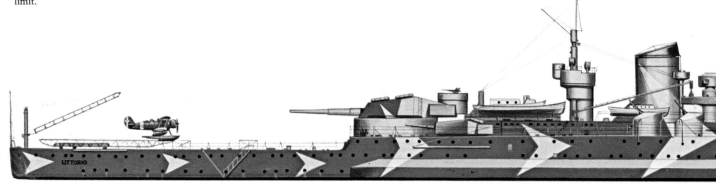

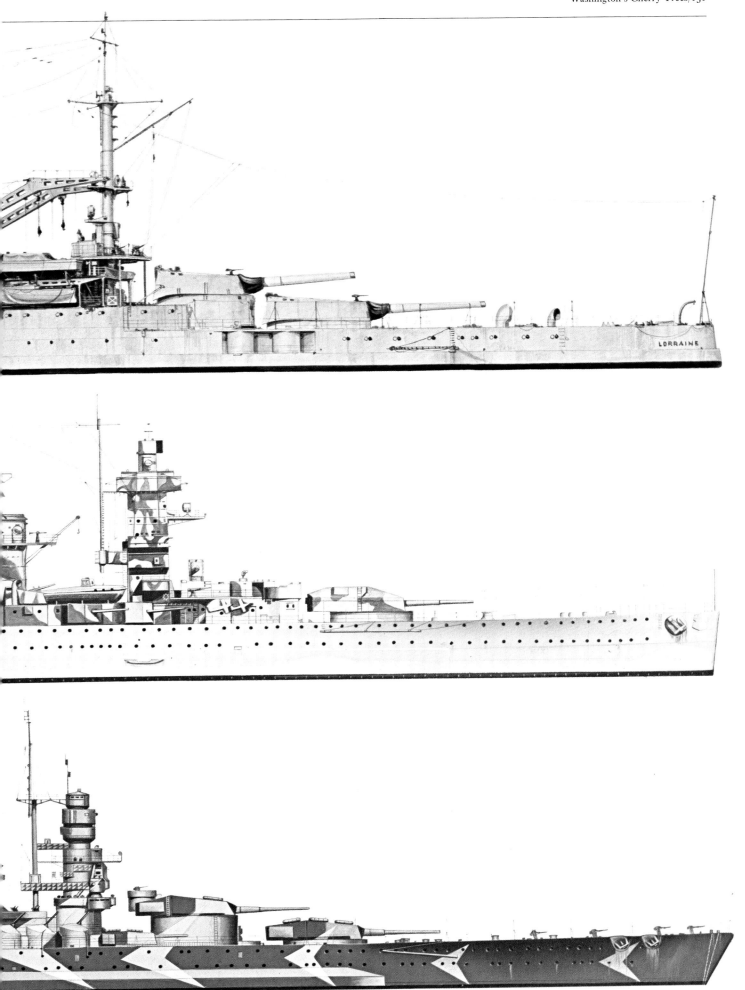

aircraft armament than their predecessors.

The British, having fought successfully for two new battleships at Washington, then did the least of all the major navies to modernize their battlefleet. Apart from trunking the funnels of the *Queen Elizabeth* Class and giving the *Repulse* a strake of 9-inch belt armor nothing was done in the 1920s, although the Admiralty was aware of the need to increase deck armor in all and in the *Queen Elizabeth* Class in particular. The problem lay in the public image of the Royal Navy as the guarantor of international peace. While other countries virtually laid up most of their battle squadrons to allow them to be modernized the British kept most of their capital ships in commission all the time. It was a symptom of the prevailing British obsession with the externals of Empire rather than the essentials, for it was deemed too risky to lay up, say, a squadron of four battleships for two or three years. There was also the dead hand of the Treasury, which had ordained that no major expenditure could be permitted until such time as a war looked likely. This was the infamous 'Ten Year Rule,' under which military and naval planners were told to assume that no major war would occur for 10 years. As a mechanism for stopping major defense expenditure in the postwar period it was successful but within a very short time it was being extended on a year-by-year basis, effectively stopping *any* future program of any significance. Other countries suffered from the world depression but the British took a perverse pleasure in planning their defense programs in a way calculated to destroy the armed forces.

Eventually, however, the restrictions were lifted, and in 1934 HMS *Warspite* was taken in hand for reconstruction. It followed the lines of other navies' modernizations, using the weight and space saved on lighter machinery and boilers to provide better protection against bombs and torpedoes. In one respect the British went a step further; because their ships

frequently had to operate in rough weather they insisted that the *Warspite*'s two catapult floatplanes must be housed in a hangar. Her two sisters *Queen Elizabeth* and *Valiant* underwent a similar reconstruction three years later and the *Renown* was given a comprehensive reconstruction to enable her to act as a fast carrier escort. Lack of money and above all time prevented the rest of the *Queen Elizabeth*s and the *Repulse* from having more than partial modernization but the ships which had not had the elevation of their guns increased to 30 degrees were given a new type of 15-inch shell with a supercharge to give them more range.

The German navy had been reduced to a coast-defense force at Versailles, with only eight pre-dreadnoughts and a collection of small cruisers and torpedo boats, but under the treaty replacements could be built. The framers of the treaty in 1919 had labored hard to prevent Germany from ever having a battlefleet again, making the limit 10,000 tons and 11-inch guns. Replacements for the *Elsass* and *Preussen* had to be ordered in 1930 and at first these were to be no more than up-to-date versions, but neither the German armed forces nor public opinion had ever accepted the restrictions of Versailles willingly. The designers promised that they could produce a

Above: The *Colorado* berthing with the aid of tugs, May 1927.

much more powerful ship with a good turn of speed and high endurance, capable of operating on the high seas as a commerce-raider. The secret was diesel propulsion and extensive use of welding, and with a speed of 26 knots and six 11-inch guns they might well live up to Fisher's dictum of ships that could run away when they were not strong enough to fight.

When the first *panzerschiff* (armored ship) *Deutschland* appeared in 1933 she caused a sensation and the English Press dubbed her a pocket battleship, a ludicrous over-valuation for what was really nothing but a heavy cruiser with 11-inch guns. Causing a sensation was just what the *Deutschland* was intended to do, for she was an expression of Germany's will to outflank the conditions of Versailles. Two more ships of similar type followed, the *Admiral Scheer* and *Admiral Graf Spee* in 1934–36, but by that time the need for purely political gestures had passed and it was realized that such ships were really too expensive for cruiser work and much too weak for the battle-line. Three more planned were cancelled, and yet foreign commentators continued to claim that they could only be matched by the three surviving British battlecruisers or the Japanese *Kongo* Class.

The completion of the *Deutschland* started the dissolution of the battleship holiday which American and British leaders had assiduously maintained, not only at Washington but at the 1930 London Naval Conference (which extended it to 1936). The French, aware of the risk to their communications with North Africa and the Far East, decided to build a new capital ship to counter the *Deutschland*, as they were allowed under the Washington Treaty. This was the 26,000-ton *Dunkerque*, which was a 30-knot battlecruiser armed with two quadruple 13-inch gun mountings. When the Germans made it clear that they were proceeding with two more *panzerschiffe* the French then laid down a second battlecruiser, the *Strasbourg*. They were heavily influenced by the *Nelson* design, and saved weight by having both quadruple turrets forward of the superstructure, but all similarity ended there for the French ships had a long flared forecastle and were among the most handsome capital ships ever built.

France had refused to ratify the 1930 treaty to leave her free to counter Germany's plans and inevitably Italy had done likewise. To match the *Dunkerque* and *Strasbourg* they announced that they would be building two 35,000-ton ships as permitted at Washington. But these turned out to be 15-inch gunned ships, the famous *Littorio* and *Vittorio Veneto*. They incorporated the same ideas as the reconstructed ships, the Pugliese system of underwater protection and similar arrangements for the fire control and bridgework, and like all Italian ships they were as fast as possible – 30 knots on trials. Unlike the old battleships they were heavily armored, but at the price of exceeding the treaty limit by 6000 tons (17 percent), a fact which the Italian navy omitted to publicize.

The French were now rightly suspicious of Italian intentions and announced that they would build two 15-inch ships to give them parity, and the Mediterranean arms race was well under way. The disarmament treaties were clearly crumbling, and the election of Adolf Hitler in 1933 speeded up the process. The new Kriegsmarine cancelled the last three *panzerschiffe* and used their guns to arm two new, nominally 26,000-ton, battlecruisers, the *Scharnhorst* and *Gneisenau*. In keeping with the previous practice of understating tonnage to evade the Versailles restrictions the new ships actually displaced 32,000 tons, an 'error' of 23 percent. Even the order for the *Gneisenau* was kept secret until Hitler publicly and unilaterally abrogated the Versailles Treaty, and a shocked world learned that two 35,000-tonners with 15-inch guns would be laid down

Below: The pride of Mussolini's new navy, the *Littorio* on trials. Her spindly control tower is in fact the armored communication tube and the main structure has not yet been added.

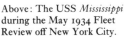

Above: The USS *Mississippi* during the May 1934 Fleet Review off New York City.

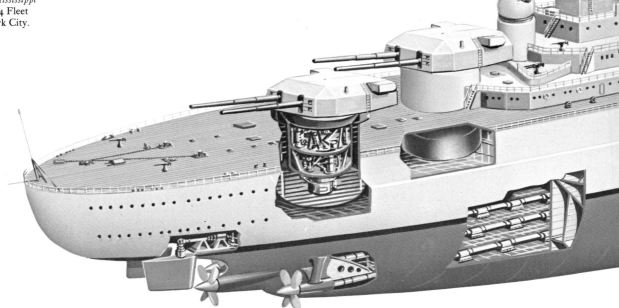

in 1935–36. They would eventually appear as *Bismarck* and *Tirpitz* and like the battlecruisers they would exceed their stated displacement by 20 percent.

As early as 1934 the British Admiralty started to plan for a war against Germany with the likely addition of Italy. A war with Japan was also considered likely, with or without Germany as an ally of the Japanese. With German rearmament now in full swing it would be necessary to start new battleships as soon as the current treaty expired in 1936, for it was reckoned that war would break out by 1941 at the latest. Design work for these '1936 Battleships' started immediately, with certain clear-cut directives. Speed was less important than armor, because it was felt that the decisive range would be 12,000–16,000 yards. High speed (30 knots) might enable longer ranges to be chosen but all experience in 1914–1918 had shown that destruction only occurred at the

lower ranges. Gun-caliber was not important but it was essential to provide protection against 16-inch shellfire.

Both the Americans and the British supported the idea of bringing gun-caliber down to 14-inch in the next treaty but as there was every indication that the Japanese would not sign the United States insisted that a clause should be inserted to allow them to revert to 16-inch if the Japanese failed to ratify by April 1937. The British stuck to their intention of building 14-inch ships, mainly because their weakened armaments industry needed as much time as possible to build and test the radically new design of quadruple turret planned. The new ships were planned to have three quadruple 14-inch turrets to provide maximum weight of broadside, and as machinery and boilers were now much lighter it would be possible to make 28 knots instead of the 23 knots originally specified. There were other

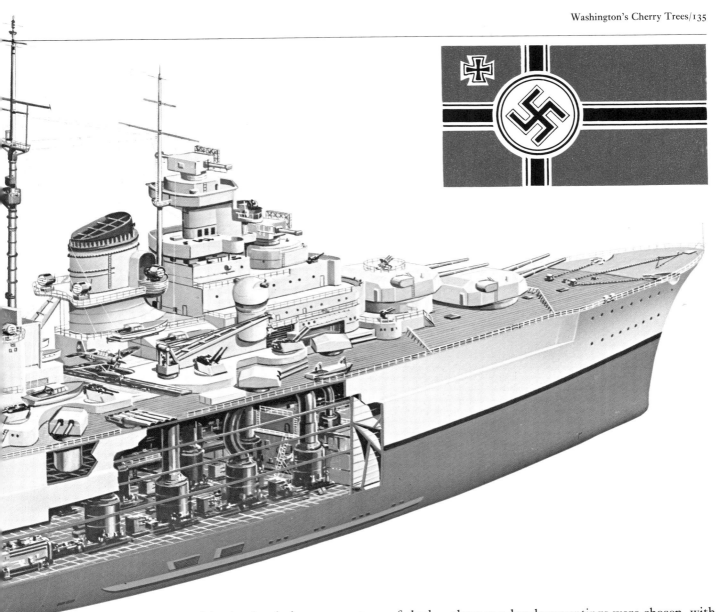

novelties in the design, a new type of dual-purpose 5.25-inch gun to avoid the weight of separate anti-destroyer and antiaircraft armaments and integral hangars for the floatplanes.

The first two of these ships were ordered in September 1936, the *King George V* and *Prince of Wales*, and it was hoped to have them at sea in 1940. At the last minute the armament was changed to ten 14-inch by substituting a twin turret for B quadruple mounting, a minor change which threw a major spanner in the works, for it delayed the order for the main armament. Three more ships were ordered under the 1937 program, and although the Admiralty had hoped to give them 16-inch guns it was realized that there was no time for such luxuries and the *Duke of York*, *Anson* and *Howe* were repeats of the *King George V*. As a class they attracted a lot of criticism, but this stemmed mainly from the alleged weakness of their main armament; they were in fact very well protected and in battle the 14-inch did all that was required of it.

Oddly enough the American solution to their requirements for a '1937 Battleship' took much the same form. Initially twelve 14-inch in

three quadruple mountings were chosen, with 28 knots speed as the most that could be achieved without sacrificing range and protection. Where the American design differed from the British was in having protection against 14-inch shellfire rather than 16-inch, for the US Navy demanded much more endurance. Like the British they chose dual-purpose secondary guns, twin 5-inch, but did not provide hangars for the floatplanes. Having an extra year in hand and much greater resources the Americans could afford to delay the order for the 14-inch guns, and when the Japanese predictably refused to ratify the 1936 London Naval Treaty triple 16-inch guns were immediately substituted for the 14-inch. The *North Carolina* and *Washington* were thus upgunned before being laid down, thanks to some intelligent contingency planning by the General Board.

In the next class, the four *South Dakota*s, the designers were set a much harder problem for they had to provide protection against 16-inch shellfire, the same speed and armament as the *North Carolina* but still keep within the 35,000-ton limit. The only area in which economies could be achieved was propulsion, and the designers managed eventually to squeeze 130,000 shaft horsepower into less space than the *North Carolina* needed for 120,000 shp. Like all the later treaty battleships the *South Dakota*s exceeded the limit in fighting trim but this would have been corrected by such subterfuges as not embarking the full outfit of ammunition in peacetime. Similarly the *King George V* Class finally grew to 38,000 tons, but as World War II had broken out by the time they were completed nobody cared.

Following their refusal to sign the 1936 Treaty the Japanese withdrew behind a veil of secrecy, although avowing their willingness to abide by the spirit of the existing treaties. What this meant in practice was that they immediately started to plan for a squadron of enormous ships which would be built in secret and then revealed when the Japanese government was ready to denounce all treaty limitations. The reasoning behind this decision was tortuous. First of all, the latest treaty expired in 1940 so the Japanese reckoned that the sudden appearance of 63,000-ton ships shortly after the expiry date was not cheating! It was also reckoned that if these ships were made big enough any battleship the Americans could build in reply would be too big to pass through the Panama Canal. It was assumed that the Americans would either have to face defeat in battle or meekly concede mastery of the Pacific to Japan. With this Byzantine logic the Japanese navy took its first steps down the road to Pearl Harbor.

The ships which the Japanese Bureau of Naval Construction designed were the ultimate in battleships, displacing 69,500 tons, armed with nine 18-inch guns and steaming at 31 knots. After much consideration it was agreed

by the Naval Staff that they were too lightly protected and lacked endurance, so size came down to 64,000 tons and diesels were substituted to give a speed of 27 knots. Then in July 1936 came an unpleasant shock; the new high-speed diesel was a failure and it was necessary to revert to four-shaft steam turbines. Two ships were laid down, the *Yamato* and *Musashi*, while a third unit, the *Shinano* followed later. Everything about them was on a gigantic scale: 16.1-inch armor on the belt, 7.75-inch deck armor, 157-ton guns capable of firing a 3220-pound shell a distance of 45,000 yards. Each triple turret weighed 2730 tons and was protected by 24.8-inch face armor and because of the 100 pounds per square inch blast from the 18-inch (15 pounds blast can knock a man out) the entire antiaircraft armament was protected by enclosed mountings. Building the first two put an almost intolerable strain on Japanese heavy industry; it was necessary to spend $10,000,000 on new plant to make the 68.5-ton armor plates and a special transport had to be built to carry the 18-inch guns and mountings from the Kamegakubi works to the shipyards.

The keel of the *Yamato* was laid down in November 1937 and *Musashi* followed in March 1938, the former at Kure and the latter at Nagasaki. They were launched in total secrecy, the *Yamato* being hidden from view by a gigantic roof at the landward end of the building dock and *Musashi* by 408 tons of camouflage netting. Not even the German naval attaché, Admiral Wenneke, was allowed to know the details of the design, but his short visit to the building berth gave him enough idea of the size to enable the Germans to make a remarkably shrewd guess at her configuration and characteristics.

Below: The Soviet *Marat* with her 'hammer funnel and sickle bow' at the 1937 Coronation Naval Review in Spithead. This was the first state visit of a Russian warship to England since 1911.

It remains only to describe two more classes of super-battleships. Germany planned six 56,000-tonners to follow the *Bismarck* and her sister *Tirpitz*, armed with eight 16-inch and driven by diesels to give them maximum range as commerce-raiders. Known only as *H* and *J* the first two were started in 1939 and were planned for completion in 1944, the year in which Hitler told the navy he meant to go to war. The Soviets, long excluded from everyone's calculations, were planning their own fleet, having first asked American and Italian firms to tender for 16-inch guns, armor plate and even complete designs. With all the various proposals in their hands the Soviets proceeded to take what they felt to be the best features and came up with a 60,000-ton ship armed with nine 16-inch and steaming at nearly 30 knots. They would have been nearly as long as the *Yamato* and almost the same

beam, although they carried nothing like the same scale of armor. Two were laid down in 1938, one at Leningrad and the other at Nikolaiev on the Black Sea.

This second naval race showed how far naval technology had advanced since 1914. Not only had size and gunpower shown a spectacular increase but propulsion had improved out of all recognition. As a simple example a *King George V* Class battleship developed 110,000 horsepower on less weight than the *Invincible* had needed for 44,000 horsepower in 1908. The Washington and London Treaties might not have averted war but they had a beneficial effect on battleship design by forcing engineers to use new steels and save weight. But all the advances between 1919 and 1939 were soon to be completely overshadowed by wartime developments. In the process the battleship would go through her final changes.

8. THE EUROPEAN WAR

When Winston Churchill said, 'We are fighting this war with the ships of the last,' he was speaking no more than the truth. Fortunately for the Royal Navy there was no High Seas Fleet across the North Sea, only the nucleus of the big fleet that Admiral Raeder had wanted to build. The British could look forward to a pair of new battleships in 1940–41 to match the *Bismarck* and *Tirpitz* and three more a year after, whereas the German program was already running late, with the *Bismarck* unlikely to be ready until early 1941. Even if Italy did come into the war the British knew that the French Fleet was more than equal to the challenge of keeping the Mediterranean under Allied control.

The British followed the strategy that had proved so successful in World War I, moving the Home Fleet to Scapa Flow to block the exits to the Atlantic, but this time the great fleet base did not prove immune to attack. On the night of 13–14 October 1939 *U-47* under Kapitänleutnant Günther Prien penetrated the line of blockships in Kirk Sound and found the old battleship *Royal Oak* lying at anchor. Prien's first salvo of three torpedoes, fired from ahead, failed to do any damage for the one torpedo which hit apparently struck an anchor cable or detonated only partially. After an interval to reload Prien fired another salvo and this time there was a loud explosion underneath the *Royal Oak* and she rolled over and sank 13 minutes later. Several books try to give the impression that she was destroyed by a colossal explosion but Admiralty records show only that the ship capsized and sank after heavy flooding amidships. It was noted that the loss of 24 officers and 809 men was much larger than it need have been because the ship's company had not yet become accustomed to war-routine. No doubt feeling secure in a quiet corner of the anchorage the ship's officers had not ensured that all the watertight doors and hatches were closed and so the ship was at her most vulnerable.

The *Royal Oak* was only fit for second-line duties but the skill of *U-47* in penetrating a base which had been thought impregnable shook the confidence of the Admiralty. As in 1914 the Home Fleet was sent away to bases on the west coast of Scotland until the defenses

could be strengthened. During this time the *Nelson* was damaged by a magnetic mine while entering Loch Ewe and the *Barham* was damaged by a torpedo but the Kriegsmarine was not able to take advantage of the Home Fleet's weakness. Not until the invasion of Norway in April 1940 did the respective heavy units come into contact. The *Renown* narrowly missed intercepting the German invasion forces on 6 April but three days later in very bad weather she surprised the *Scharnhorst* and *Gneisenau* some 50 miles from Vestfjord. The British ship's lookouts spotted the Germans through the snow squalls and closed to within nine miles before opening fire. The Germans recovered quickly from their surprise but the *Renown*'s shooting was better and she put *Gneisenau*'s forward fire control out of action before the German ships used their superior speed to get away in the murk. Hitler's orders were quite specific that capital ships were not to expose themselves to damage and so Admiral Lütjens felt obliged to break off the action – the German navy was being handled as delicately as ever.

In the Second Battle of Narvik the veteran *Warspite* distinguished herself by following a force of nine destroyers up Ofotfjord to hunt down a force of German destroyers. It was hardly an ideal setting for a battleship, with a risk of grounding or being torpedoed but Admiral Whitworth's gamble paid off. *Warspite*'s Swordfish floatplane was able to reconnoiter for the whole force and her 15-inch guns completed the destruction wrought by the destroyers' guns and torpedoes. At the end of the day the entire German force of eight destroyers had been wiped out, and the *Warspite* had begun a career of extraordinary good fortune.

Norway showed that air attack was dangerous to battleships but it also showed

Above: HMS *Warspite*, flagship of the Mediterranean Fleet and the most famous British battleship of World War II.

how fallacious had been the claims of people like Billy Mitchell in the 1920s. The USAAF and the RAF had, largely as a result of Mitchell's highly dramatized stunt demonstration in 1921, put their faith in high-level bombing, but again and again it was to prove inaccurate against ships at sea. What was deadly was dive-bombing, and when the *Rodney* was damaged and several smaller ships sunk by Ju 87 Stuka dive-bombers it was realized that gunfire alone could not defend ships. Multiple light guns like the 2-pounder pompom (known as the 'Chicago Piano' from James Cagney films) broke up massed attacks and heavier guns could force level-flight bombers to keep high but fire control was not yet sufficiently advanced to ensure more than random hits. The vulnerability of aircraft carriers was also demonstrated when, during the closing stages of the campaign, the *Scharnhorst* and *Gneisenau* caught HMS *Glorious* after she had evacuated the last RAF aircraft and aircrew from Norway. The carrier was unable to launch her Swordfish torpedo-

bombers and so the two battlecruisers carried out a leisurely target-practice. The only opposition came from the destroyer *Acasta*, which managed to hit *Scharnhorst* with a torpedo before being sunk.

The next action involving capital ships was the tragic destruction of the French fleet at Mers-el-Kebir in July 1940. After the fall of France the old battleships *Paris* and *Courbet* had escaped to England and the incomplete *Richelieu* and *Jean Bart* managed to reach North Africa, but four of the remaining capital ships, the *Dunkerque*, *Strasbourg*, *Bretagne* and *Provence* had been instructed by the new Pétain Government to remain at Mers-el-Kebir near Oran, under the terms of the armistice negotiated with Hitler. The British were understandably alarmed at the collapse of the joint strategy in the Mediterranean, for Italy had chosen this moment to attack France. To guard against any Italian move the Admiralty immediately formed Force H under Admiral Somerville, including initially the *Hood*, *Valiant* and *Resolution* and the new carrier *Ark Royal*.

The British were particularly worried that somehow the Italians or the Germans might be able to seize the French ships. The Italians were close enough to make such a move, and even if the armistice kept the Germans out of southern France there was the chance that they might go through Spain to capture Gibraltar.

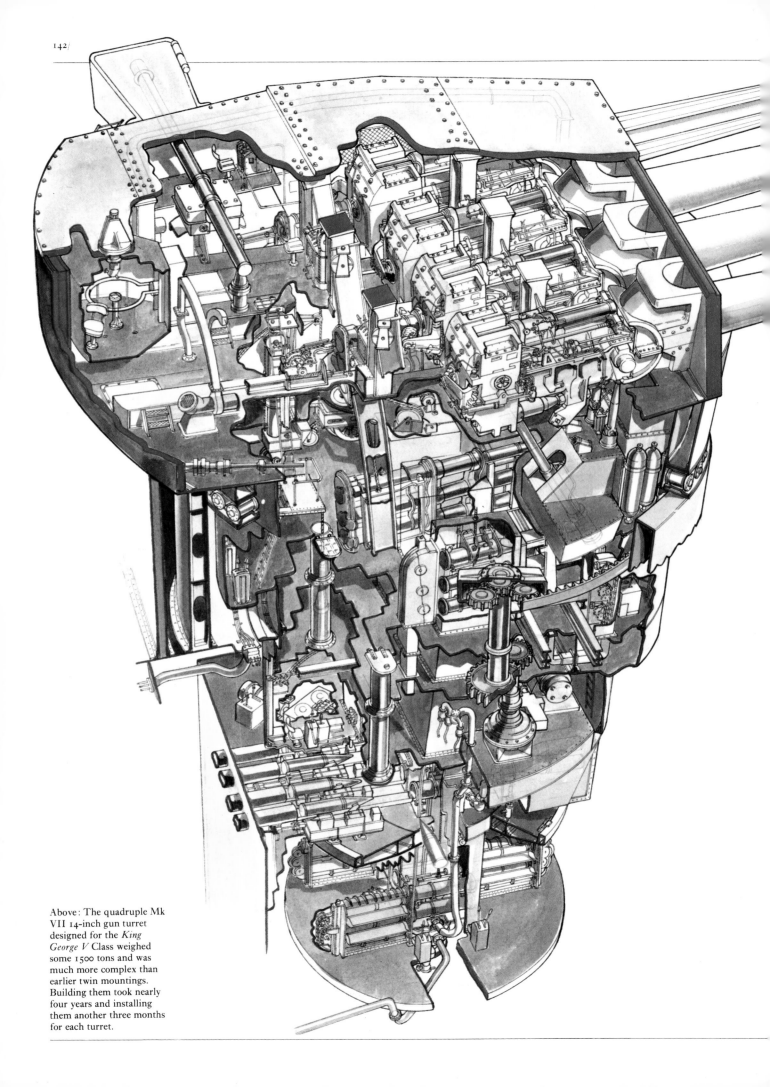

Above: The quadruple Mk VII 14-inch gun turret designed for the *King George V* Class weighed some 1500 tons and was much more complex than earlier twin mountings. Building them took nearly four years and installing them another three months for each turret.

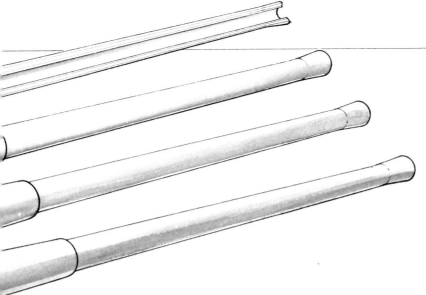

Neither of the Axis partners' guarantees about keeping their hands off the French fleet could be trusted and so the British felt compelled to take action. Somerville was ordered to make an offer to the French Commander in Chief at Mers-el-Kebir, Admiral Gensoul, and if necessary to back it up by force.

The five alternatives given to Gensoul were: to put to sea and join the Royal Navy to continue the fight; to accept internment in a British port; to be demilitarized in a French West Indian port; to scuttle the ships; or to demilitarize them at Mers-el-Kebir. They put Gensoul in an almost impossible position: the first two conditions were a direct contravention of the armistice, and even the third could be construed as a contravention; the fourth was an insult to the honour of the French navy and the fifth was impossible to achieve in the six hours allowed. Even if the French navy could accept the conditions the Pétain Government would never have allowed it as the fleet was virtually the last bargaining counter that it had, and so its instructions to the Commander in Chief Admiral Darlan were to reject the ultimatum. To make matters worse Somerville and Gensoul do not appear to have been the

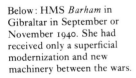

Below: HMS *Barham* in Gibraltar in September or November 1940. She had received only a superficial modernization and new machinery between the wars.

right sort of people to conduct such delicate negotiations. Inexplicably Gensoul reported back to Darlan that he had been ordered to scuttle his ships or surrender them. The other alternatives were not explained and naturally Darlan ordered him to resist.

The result was that on 3 July at 1745 hours Force H opened fire on the crowded harbor. The 15-inch salvos quickly overwhelmed the French and the *Bretagne* blew up, while the *Dunkerque* and *Provence* were badly damaged. The only large ship to escape the carnage was the *Strasbourg*, which got clear behind dense clouds of smoke. The *Hood* was only capable of 28 knots and could not catch the French battlecruiser, but Somerville had carried out his orders and the bulk of the French fleet was now immobilized. Of course as a consequence French opinion was outraged and the French navy in particular became violently anti-British. Any hope that individual ships might desert the Vichy cause to join General de Gaulle's Free French forces in England was dashed. Nor did the failure to take Dakar in September 1940 improve relations, for the *Richelieu* was damaged twice by British attacks. This did not stop her from firing on the attacking force with her 15-inch guns, and when HMS *Resolution* was damaged by a torpedo from the submarine *Bévéziers* British and Free French enthusiasm evaporated.

For a while it looked as if the powerful Italian fleet might force the British to abandon the Mediterranean but this gloomy view took no account of the difference in temperament between the British and the Italians. Admiral Cunningham was given the *Warspite* as his flagship and the unmodernized *Royal Sovereign*, *Ramillies* and *Malaya* as well as the old carrier *Eagle*. With this force Cunningham

had no hesitation in giving battle to the Italians, and the result of the action off Calabria on 9 July 1940 showed that his confidence was justified. Both the Italians and the British were at sea to cover the passage of convoys when Admiral Campioni's squadron was sighted by the British. A torpedo-attack by the *Eagle's* Swordfish was unsuccessful but the *Warspite* was at extreme gun-range. As the Italian ships worked up to full speed, heading for the horizon, the *Warspite's* guns fired their first salvos against an enemy battleship since Jutland 24 years earlier. Suddenly an orange flash appeared on the flagship *Giulio Cesare* as a 15-inch shell landed alongside her funnel. That was the end of the action for the Italians retired under cover of a smokescreen and used their higher speed to get away. The British were naturally disappointed but the action showed that they had little to fear from the Italian fleet. Incidentally, the *Warspite's* record of a hit at 26,400 yards still stands, the greatest range at which guns have scored a hit on a moving target at sea.

The next action, between Force H and Campioni off Cape Spartivento on 25 November, was inconclusive but Cunningham had already inflicted a serious defeat on the Italians. With a new carrier, HMS *Illustrious* and another modernized battleship, the *Valiant*, he was able to attack the enemy in his main base at Taranto. On the night of 11–12 November a force of 21 Swordfish biplane torpedo-bombers attacked Taranto, sinking the *Conte*

di Cavour and severely damaging the new *Littorio* and the older *Duilio*. Although the damaged ships would be repaired the attack meant that three out of six Italian battleships were out of action. More important, the Italians had lost the initiative to the British, who could now send convoys through the Mediterranean to reinforce their units in Malta and Egypt. Ever since 1918, when it had planned to use Sopwith Cuckoo torpedo-bombers to attack the High Seas Fleet, the Fleet Air Arm had hoped for such an opportunity, and now it had made history. Among the interested parties who studied the results of Taranto were the Japanese, who were at that moment thinking of ways to do much the same thing to the Americans at Pearl Harbor.

Early in 1941 Cunningham's battleships at last had a real chance to strike a blow at the Italian fleet. On 28 March the carrier *Formidable's* torpedo-bombers put two torpedoes into the *Vittorio Veneto*, giving the Mediterranean Fleet a chance to cut her off. At dusk Cunningham's ships were still 65 miles astern but he had already made up his mind to risk a night action and pressed on. Since Jutland the Royal Navy had learned a lot and although only a few ships had radar they were equipped with such aids as flashless cordite and the drills had been practised endlessly. In contrast the Italians were badly equipped, with no radar but above all lacking any sort of training for night-fighting.

The object of the pursuit, the *Vittorio Veneto*

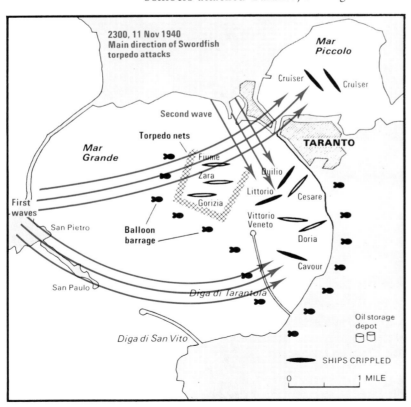

Left: The *Littorio* maneuvering under full helm.

had actually escaped and was heading for home but just before dusk a stray aircraft torpedo had hit one of her escorting cruisers, the *Pola*. Admiral Iachino ordered two of her sisters, the *Fiume* and the *Zara* to go back and try to tow her home. It was while these two ships were looking for their stricken consort that they appeared as echoes on the British radar screens. Suddenly they realized that they had fallen into a trap, but too late. In Cunningham's own words,

'In the dead silence, a silence that could almost be felt, one heard only the voices of the gun control personnel putting the guns on the new target. One heard the orders repeated in the Director Tower behind and above the bridge. Looking forward, one saw the turrets swing and steady when the 15-inch guns pointed at the enemy cruisers. Never in the whole of my life have I experienced a more thrilling moment than when I heard a calm voice from the Director Tower: "Director-layer sees the target,"

sure sign that the guns were ready and that his finger was itching on the trigger. The enemy was at a range of no more than 3800 yards – point blank.

It must have been the Fleet Gunnery Officer . . . who gave the signal to open fire. One heard the "ting-ting-ting" of the firing gongs. Then came the great orange flash and the violent shudder as the six big guns bearing were fired simultaneously. At the very same moment the destroyer *Greyhound* . . . switched her searchlight on to one of the enemy cruisers, showing her up momentarily as a silvery-blue shape in the darkness. Our searchlights shone out with the first salvo and provided illumination for what was a ghastly sight. Full in the beam I saw our six great projectiles flying through the air. Five out of the six hit a few feet below the level of the cruiser's upper deck and burst with flashes of brilliant flame. The Italians were quite unprepared. Their guns were trained fore and aft. They were hopelessly shattered

Below: The *Andrea Doria* and *Giulio Cesare* at speed during an action with British cruisers and destroyers in December 1941.

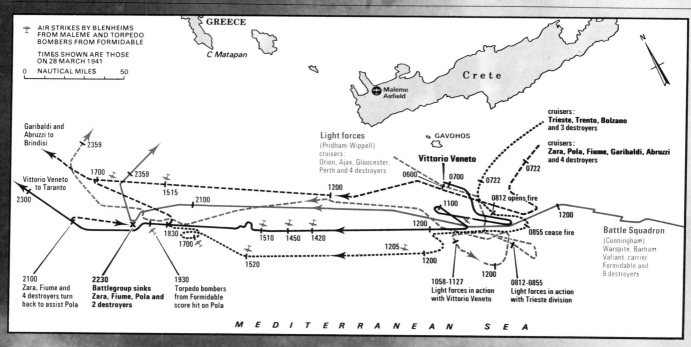

AIR STRIKES BY BLENHEIMS
FROM MALEME AND TORPEDO
BOMBERS FROM FORMIDABLE

TIMES SHOWN ARE THOSE
ON 28 MARCH 1941

0 NAUTICAL MILES 50

GREECE

C Matapan

Crete

Maleme
Airfield

GAVDHOS

cruisers:
Trieste, Trento, Bolzano
and 3 destroyers

cruisers:
Zara, Pola, Fiume, Garibaldi, Abruzzi
and 4 destroyers

Light forces
(Pridham-Wippell)
cruisers:
Orion, Ajax, Gloucester,
Perth and 4 destroyers

Vittorio Veneto

0600

0700

0722

0812 opens fire

1100

0855 cease fire

1200

1200

1200

Battle Squadron
(Cunningham)
Warspite, Barham,
Valiant, carrier
Formidable and
9 destroyers

Garibaldi and
Abruzzi to
Brindisi

2359

1700

2359

1515

2100

1200

Vittorio Veneto
to Taranto

2300

1830

1700

1510 1450 1420

1205

1520

1200

1200

2100
Zara, Fiume and
4 destroyers turn
back to assist Pola

**2230
Battlegroup sinks
Zara, Fiume, Pola and
2 destroyers**

1930
Torpedo bombers
from Formidable
score hit on Pola

1058-1127
Light forces in action
with Vittorio Veneto

0812-0855
Light forces in action
with Trieste division

M E D I T E R R A N E A N   S E A

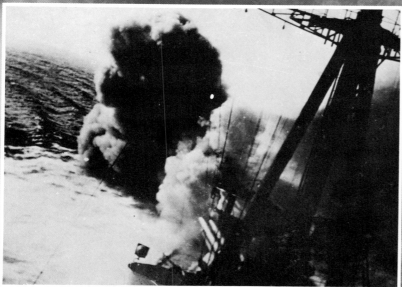

Left: An Italian battleship
firing at British cruisers
during the inconclusive
Battle of Cape Spartivento in
November 1940.

Below: The burning hulk of the *Admiral Graf Spee* off Montevideo in December 1939. The wreckage at far left is the after 11-inch turret blown off its barbette. The funnel has collapsed to port.

before they could put up any resistance.'

The Battle of Cape Matapan, or Gaudo Island to the Italians, was a massacre. The *Warspite*, *Barham* and *Valiant* sank the *Fiume* and *Zara* in minutes, despite a gallant attempt by the escorting destroyers to defend them. Then the British destroyers hunted down the crippled *Pola* and torpedoed her after taking off as many survivors as they could.

Cunningham did not regard Matapan as anything more than a skirmish for he had intended to sink the *Vittorio Veneto* but its strategic value came to light two months later. During the evacuation of Crete the Mediterranean Fleet was exposed to merciless dive-bombing and was suffering heavy losses. Now was the time for the Italian fleet to put to sea but somehow it had better things to do, and its opportunity to avenge Matapan passed. Even after German U-Boats sank the carrier *Ark Royal* and HMS *Barham* and Italian frogmen had put the *Queen Elizabeth* and *Valiant* out of action in Alexandria the memory of Matapan continued to inhibit the Italian high command, and they lost their last chance to dominate the Mediterranean.

Back in Home Waters the British faced a serious threat from the German heavy surface ships. The sinking of the *Admiral Graf Spee* at

the River Plate had demolished the reputation of the *panzerschiff* but in 1939–40 and 1941 the *Scharnhorst* and *Gneisenau* had sortied into the North Atlantic, sinking the armed merchant cruiser HMS *Rawalpindi* and 22 merchant ships totalling over 115,000 tons gross. To counter the threat to the Atlantic convoys the Admiralty reinforced each convoy with an old battleship. Although these veterans were totally outclassed, on the three occasions that the German ships made contact the sight of a tripod mast and control top was sufficient to make them sheer off. Hitler's orders had to be obeyed, even when they ran counter to common sense.

A much more dangerous threat was the battleship *Bismarck*, which completed her training and shakedown by April 1941. With a margin of 6000 tons over the Washington Treaty limit the designers had been able to produce a balanced design, fast, well armed and well protected. But in spite of claims about massive armor of a mysterious new type it was in many ways a conservative design, based on the old *Baden*, which the Germans regarded as their most successful battleship. Thus there was no attempt to provide a dual-purpose secondary armament and the armor scheme emphasised defense against long-range gunfire rather than bombs. Nor was there any great measure of vertical 'sandwich' protection as in the latest American and British ships. She was fast but mainly because of a long hull, and the machinery spaces were remarkably large by other navies' standards. The armament was very similar to the *Baden* Class, eight 15-inch (38 cm) in four twin turrets, widely regarded as the best possible arrangement. To describe her as a reworked design is not to belittle the *Bismarck* or her designers, for the *Baden* had been an excellent ship. Furthermore, given the haste with which Hitler wanted the German navy expanded there was no time to spend on elaborate testing of alternative ideas – for much the same reason the *Scharnhorst* had drawn on the 1914 design of the *Mackensen* Class.

When the Admiralty learned of the movement of two large ships westward out of the Baltic it was obvious that the *Bismarck* and the new heavy cruiser *Prinz Eugen* were ready to break out into the Atlantic. On 21 May they sailed from Bergen and disappeared but already two heavy cruisers, HMS *Norfolk* and HMS *Suffolk* were patrolling their most likely exit-route, the Denmark Strait between Iceland and Greenland. As the *Suffolk* was equipped with radar she was able to keep in contact in spite of the bad visibility, and on 23 May the Battlecruiser Squadron, comprising HMS *Hood* and the new fast battleship *Prince of Wales* sailed from Scapa Flow.

Vice-Admiral Holland in the *Hood* hoped to intercept the German ships on a bearing which would give his ships the benefit of their heavier gunpower but an alteration of course during the night meant that his squadron came upon the enemy while they were on a fine bearing. This meant that only A and B turrets could fire, and as one of *Prince of Wales'* forward guns could fire only once before jamming the British were facing eight 15-inch and eight 8-inch with four 15-inch and five 14-inch guns. The *Prince of Wales* had been completed only two weeks earlier; her gun turrets still needed adjustments and the crew was young and relatively untrained. Holland also knew that his flagship's immunity zone (i.e. the range at which her armor could not be pierced) extended from 18,000 yards down to 12,000 yards, whereas the *Prince of Wales* was immune out to maximum range. Prudence might dictate that the well-protected ship should be sent first but Holland could hardly expect to inspire confidence by leading from behind, and it would have been rash to expose such a raw ship to the full fury of German fire. All that he could do was close the range as fast as he could until the *Hood* was inside her immunity zone and then turn to open the arcs of fire of the two ships' after turrets. Another factor weighing on Holland's mind was undoubtedly the difference in speeds; the *Hood* was capable of no more than 28.5 knots whereas the German ships were credited with 31 knots.

The action opened briskly with *Hood* firing

well on ranges supplied by her Type 284 radar. Her first three salvos were right for distance but off line, and in the opinion of the *Bismarck*'s surviving 3rd gunnery officer the next salvo looked likely to hit. But instead, just as the *Hood* turned to port to bring her full broadside to bear she vanished in a huge explosion. When the smoke cleared the two halves of the ship could be seen disappearing, just like the battle-cruisers at Jutland. Then a tornado of fire burst about the *Prince of Wales* as both *Bismarck* and *Prinz Eugen* concentrated their fire on her. She was hit seven times, by four 15-inch and three 8-inch shells, the worst being a 15-inch shell which passed through the compass platform without detonating. It scattered fragments of the binnacle around, killing or wounding everyone except Captain Leach.

Fortunately the other six hits were not as deadly, some detonating partially and others not at all. The main gunnery radar set was not working but the Type 281 air-warning set was capable of passing ranges to the guns, and the *Prince of Wales* fought back gamely. The inexperienced crew of X turret inadvertently let a 14-inch shell drop out of the hoist, and it rolled into the roller-path and jammed the turret. With a great deal of bad language and musclepower the 1400-pound projectile was manhandled back into the hoist and the turret resumed firing. Straddles were obtained, and it was later learned that two 14-inch shells hit the *Bismarck* below the waterline, a creditable achievement for any new ship.

To the surviving senior officer, the Rear Admiral commanding the two heavy cruisers, it looked very bad for the *Prince of Wales*, and he ordered Leach to break off the action. Perhaps he was right, for it was dangerous to expose such a valuable ship to any further

Below: HMS *Nelson* down by the bow after being hit by an Italian aerial torpedo while escorting a Malta convoy.

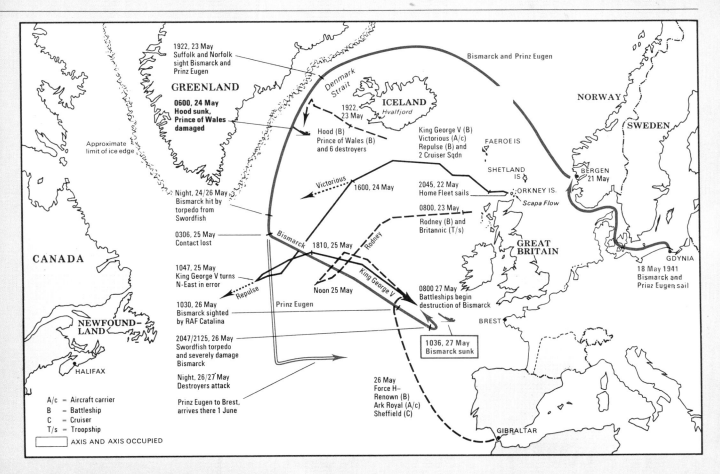

1922, 23 May
Suffolk and Norfolk
sight Bismarck and
Prinz Eugen

GREENLAND

**0600, 24 May
Hood sunk,
Prince of Wales
damaged**

Bismarck and Prinz Eugen

Denmark Strait

ICELAND
Hvalfjord

1922,
23 May

Hood (B)
Prince of Wales (B)
and 6 destroyers

King George V (B)
Victorious (A/c)
Repulse (B) and
2 Cruiser Sqdn

NORWAY

SWEDEN

FAEROE IS

SHETLAND
IS

BERGEN
21 May

ORKNEY IS.

Approximate
limit of ice edge

Victorious

1600, 24 May

2045, 22 May
Home Fleet sails

Scapa Flow

Night, 24/26 May
Bismarck hit by
torpedo from
Swordfish

0800, 23 May
Rodney (B) and
Britannic (T/s)

0306, 25 May
Contact lost

Bismarck

1810, 25 May

Rodney

King George V

GREAT
BRITAIN

GDYNIA

CANADA

1047, 25 May
King George V turns
N-East in error

Repulse

Noon 25 May

Prinz Eugen

0800 27 May
Battleships begin
destruction of Bismarck

18 May 1941
Bismarck and
Prinz Eugen sail

1030, 26 May
Bismarck sighted
by RAF Catalina

BREST

1036, 27 May
Bismarck sunk

NEWFOUND-
LAND

2047/2125, 26 May
Swordfish torpedo
and severely damage
Bismarck

HALIFAX

Night, 26/27 May
Destroyers attack

26 May
Force H–
Renown (B)
Ark Royal (A/c)
Sheffield (C)

A/c = Aircraft carrier
B = Battleship
C = Cruiser
T/s = Troopship

Prinz Eugen to Brest,
arrives there 1 June

GIBRALTAR

AXIS AND AXIS OCCUPIED

damage but afterwards Captain Leach indignantly denied that his ship was unfit for action. Looking at the number of other actions involving German capital ships one cannot help thinking that if either the *Hood* or the *Prince of Wales* had scored a damaging hit early in the fight the two German ships would have broken off the action and returned to Bergen. Be that as it may, the *Prince of Wales* had already set in motion a train of events which would destroy the *Bismarck*.

The Admiralty now mustered all its resources to hunt down the *Bismarck*. The carrier *Ark Royal* and the *Renown* had already left Gibraltar without waiting for orders and the battleship *Rodney* left the convoy she was escorting to Halifax to join the Home Fleet. The Commander in Chief Home Fleet, Admiral Tovey was at sea in the *King George V*, heading to try and intercept the German ships, and all available aircraft and ships were searching, for the heavy cruisers *Norfolk* and *Suffolk* had lost radar contact. Late on the night of 24 May the *Bismarck* had been hit by

an 18-inch aerial torpedo from one of the *Victorious'* Swordfish but this had exploded on the armor belt amidships and had not slowed her down. What nobody knew was that the two hits from the *Prince of Wales* had damaged the *Bismarck*'s fuel tanks, contaminating a large part of her oil and leaving a long slick.

Finally on 26 May a Catalina flying boat sighted the oil slick and identified the *Bismarck*. She was now heading for Brest in Brittany, for Admiral Lütjens had realized that the Atlantic sortie would be impossible without the full load of fuel. However, the nearest British heavy ships would not be able to close the distance before the *Bismarck* came under the shelter of shore-based aircraft. It was essential to slow her down before this happened and so the *Ark Royal* was ordered to fly off a torpedo-strike.

The first wave of Swordfish nearly sank HMS *Sheffield*, one of the Home Fleet cruisers in the area, but the mistake was a fortunate one for it showed that the sea was too rough for the magnetic pistols on the torpedoes. The second

Above: The twin and quadruple 14-inch turrets of the King George V trained on the beam to avoid flooding in a bad head sea.

Below: The *panzerschiff Deutschland* was the first of her class to be built. She was later renamed *Lützow*.

strike was made with the torpedoes set to 'Contact' and this time the Swordfish found the right target. In spite of a withering fire from 105-mm down to 20-mm guns the Swordfish managed to get two hits, one on the armor and one right aft. The second wrecked the steering gear and jammed the rudders, leaving the giant ship careering helplessly in circles until she was slowed down and steered on the propellers. Throughout the night the *Bismarck*'s crew worked to free the rudders for their lives depended on it; divers might have gone over the side with explosives to blow the tangled wreckage away but the rough weather put it out of the question. All the ship could do (her consort *Prinz Eugen* had been detached to Brest two days earlier) was to sell her life as dearly as she could.

Early next morning the Home Fleet hove into sight, with the battleships *King George V* (flagship) and *Rodney*, and at 0847 hours the British fired their first salvos. This time the roles were reversed, the *Bismarck* maneuvering with difficulty and her opponents operating independently of one another. The *Rodney* approached end-on so that she could rake the

Bismarck while the *King George V* fired broadside-to-broadside. The first German salvos were accurate, straddling the *Rodney* but thereafter *Bismarck*'s gunnery fell off rapidly and within half an hour she was silenced. With her low armored deck she was very resistant to medium-range gunfire, and although the vital communications and power-supplies above the armor deck were soon largely destroyed, she stayed defiantly afloat. Admiral Tovey ordered the flagship out to 14,000 yards to get more plunging hits and sent the *Rodney* in to 4000 yards to fire full broadsides at the superstructure, but still she would not sink.

By now the battleships were running short of fuel after their long chase and Tovey decided to break off the gun action, leaving the job of sinking the tortured wreck to the torpedoes of the heavy cruiser *Dorsetshire*. Three torpedoes hit the starboard side and then the *Dorsetshire* moved around to fire a fourth into the port side, and at 1036 hours the *Bismarck* rolled over and sank. In years to come her last fight would take on some of the trappings of mythology: her armor was never perforated, she was scuttled, she could have been towed into port, etc. etc. She stood up very well to severe damage but there is no factual evidence to support the claim that her armor was not pierced. Post-war tests on 12.6-inch armor plates from her sister *Tirpitz* showed that British 16-inch and 14-inch would pierce them without any difficulty at the ranges at which the action was fought. Alleged testimony to the effect that the engine rooms were running as quietly as if 'ready for the admiral's inspection' ignore the fact that the *Bismarck* was lying low in the water and steaming at five knots after half an hour. As for the scuttling, it is relevant that the only survivors the British rescued were from topside positions; no stokers or artificers were among them, and given the

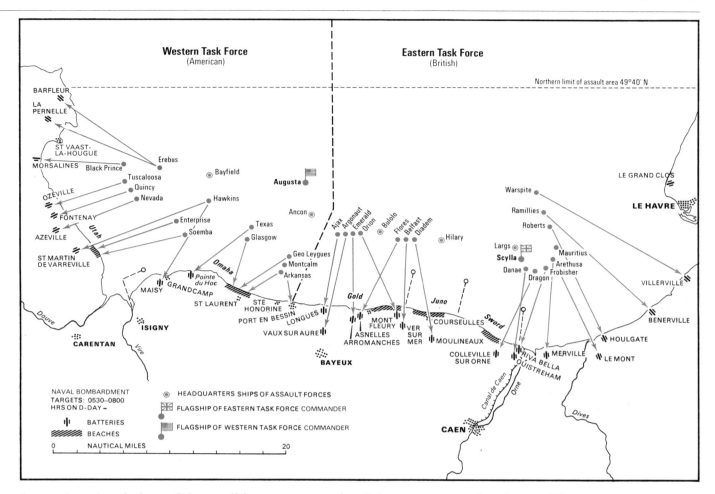

horrendous descriptions of the conditions on board this is hardly surprising.

There remains only one question to be asked. What sank the *Hood*? Naturally the Admiralty was worried about the loss of such an important ship in a manner reminiscent of the Jutland losses. Two boards of enquiry were unable to put a finger on the precise cause of loss, and with only three survivors it is impossible to produce firm evidence for any theory. However certain facts are known. First, a small fire was observed on the shelter deck among some anti-aircraft rocket-launchers but this was nowhere near a magazine. Second, both the 8-inch and 15-inch shells which hit the *Prince of Wales* did not explode properly, making it unlikely that a shell went straight into a magazine. Third, the *Hood* was at the outer edge of her immunity zone, which combined with the poor performance of German heavy shell mentioned already, makes a direct penetration of the magazine even less likely. Fourth, since 1917 both propellant and flashtight protection had been improved and thoroughly tested; no other case of spontaneous detonation of the magazines occurred in a British armored ship in World War II. There were however, two weaknesses in the *Hood*. The first was the large amount of 4-inch antiaircraft ammunition added outside the after magazines, and if this had been detonated it might have set off the main 15-inch magazines. Second, there were two pairs of above-water torpedo tubes, and if the fire set off the warheads the explosion could have broken the ship in two. The second board of enquiry heard evidence from the DNC's department to the effect that such an explosion could break the ship in two, and that seems the likeliest explanation.

The sinking of the *Bismarck* put an end to the Kriegsmarine's plans to disrupt Atlantic shipping with surface forces. Thereafter the *Scharnhorst*, *Gneisenau* and *Prinz Eugen* remained at Brest, subjected to heavy if inaccurate high-level bombing by the RAF. They were hit and temporarily put out of action several times in 1941 and so in January 1942 Hitler ordered Naval Group Command West to bring the three ships back to Germany. The plan, Operation Cerberus, had all the hallmarks of Hitlerian boldness, a daylight dash through the English Channel in the teeth of coastal guns, air attacks and surface ships. Once again Hitler had divined his opponents' weakness: nobody really *believed* that heavy units would dare to go through the Channel, and as long as a heavy umbrella of fighters was provided by the Luftwaffe the risk was small. And so it turned out on 12 February 1942, a day which *The Times* claimed was worse than any since the Dutch sailed up the Medway in 1667.

To start with, sloppy air reconnaissance failed to sight the German ships until 1042 hours by which time they were off le Havre.

Above: The US ships *South Dakota* and *Alabama* operated with the British Home Fleet in 1942 covering convoys to North Russia.

Then piecemeal attacks by MTBs and aircraft, without any attempt at coordination, were beaten off with ease. Later a small force of destroyers from Harwich miraculously got within 4000 yards but their torpedoes missed, and a torpedo strike by Fleet Air Arm Swordfish without any fighter escort was massacred. To cap it all the 15-inch guns at Dover were not allowed to open fire because the whereabouts of various friendly units was not known, and when permission to open fire was finally obtained the targets were disappearing into the haze at extreme range. But, outraged pride apart, the Channel Dash was not a victory for Hitler, and it had positive advantages for the British. Once the *Scharnhorst* and *Gneisenau* were away from Brest there was no threat of a sortie against the Atlantic convoys, and in German waters they could be contained more easily. Both ships had been damaged by mines at the end of their dramatic escape and *Scharnhorst* did not return to service until August 1942. Her sister was less fortunate, and shortly after her arrival at Kiel she was badly damaged by a bomb-hit forward. She was then towed to Gdynia (Gotenhafen) to be repaired and rearmed with three twin 15-inch turrets. This had been allowed for in the original design, and at the same time it was hoped to improve seaworthiness by lengthening the bow. Work continued into 1943 but steel and labor were in desperately short supply and work slowed to a halt in 1944. The useless hulk was finally scuttled in the harbor entrance in March 1945.

The *Scharnhorst* was sent to join the *Tirpitz* and the surviving cruisers and destroyers in northern Norway, where they could at least threaten the convoys taking supplies to Murmansk. Although the *Tirpitz* played a major part in the destruction of convoy PQ.17 it was merely a false report that she had left harbor which did the damage. For the rest of the time the remaining German heavy ships played the part of a fleet in being. When the *Tirpitz* was immobilized by British midget submarines in September 1943 it was left to the *Scharnhorst* to make some sort of effort to stop the convoys

getting through. The new Commander in Chief, Dönitz, obtained permission from Hitler to mount a major operation but on condition that Hitler refrained from interfering. Northern Group was commanded by Admiral Kummetz but he had to hand over command to Admiral Bey, who commanded Northern Group's destroyers. Bey was under the impression that a destroyer-raid was all that was intended but at the last minute he was told that the *Scharnhorst* ought to be sent out as well. By 19 December 1943 the planning was complete and Hitler was told by Dönitz that the *Scharnhorst* would attack a convoy as soon as the circumstances were favorable.

With the British reading many of the top-level signals the circumstances could hardly be favorable, but on Christmas Day the *Scharnhorst* and five destroyers sailed from Altenfjord. Their objective was convoy JW.55B, outward bound for Murmansk, for reconnaissance had failed to detect a second convoy, RA.55A, homeward bound from the Kola Inlet to Loch Ewe. The convoys were escorted by cruisers and destroyers, with a distant escort provided by the battleship *Duke of York*, the cruiser *Jamaica* and four destroyers. The Commander in Chief Home Fleet, Admiral Fraser was so certain of German intentions that he deliberately transferred some escorts from RA.55A to bring JW.55B's escort up to 14 destroyers. Admiral Bey was taking his force into a hornet's nest, and when he lost contact with his destroyers early next morning disaster became certain. Unaware that the *Duke of York* and *Jamaica* were only 200 miles away, closing at a steady 17 knots, the *Scharnhorst* continued on course for the convoy.

At 0840 hours on the 26th the cruiser *Belfast* picked up the *Scharnhorst* on radar at 30 miles, and three-quarters of an hour later she fired starshell into the Arctic gloom. Then the *Norfolk* opened fire and an 8-inch shell smashed into the *Scharnhorst*'s foretop, destroying the fire control director. Badly shaken, the battlecruiser turned away, her bulk enabling her to outstrip the cruisers in the heavy seas. Admiral Burnett, flying his flag in HMS *Belfast*, ordered the three cruisers to return to the convoy. It was feared for a time that the *Scharnhorst* would return to base but just after mid-day she reappeared and opened fire on the cruisers at 11,000 yards. The *Norfolk* was hit in an 8-inch turret and had most of her radar sets knocked out but the three cruisers fought back and forced the battlecruiser to turn away.

The *Scharnhorst* unwittingly turned in the direction of the *Duke of York* and at 1617 hours the Home Fleet flagship picked her up

on radar. When the range came down to 12,000 yards the *Duke of York* opened fire, taking the *Scharnhorst* completely by surprise. She sheered away sharply but the duel continued at ranges of 17,000–20,000 yards as *Scharnhorst* used her speed to open the range. Her gunnery improved but the two hits on the *Duke of York* went through masts without exploding. The *Duke of York*'s gunnery was excellent, and even when the *Scharnhorst* made small alterations of course to avoid the salvos radar-plotting was able to allow for them. One or more of the 14-inch hits damaged a propeller shaft but this was not enough to cripple the *Scharnhorst* and eventually Admiral Fraser ordered his destroyers to attack with torpedoes. Four destroyers delivered a coordinated port and starboard attack and scored four hits.

The *Duke of York* opened fire again at 10,400 yards and within half an hour the *Scharnhorst*'s speed was down to 5 knots. She was burning furiously and shrouded in smoke and when the *Jamaica* closed in for the *coup de grace* with torpedoes she reported that all she could see was a dull glow in the smoke. Nobody saw the end of *Scharnhorst* at around 1945 hours and only 36 survivors were found in the icy water. She had gone down with nearly 2000 men on board, a victim of an outdated strategy and poor intelligence.

With the *Scharnhorst* gone it remained only to account for the *Tirpitz*, which had survived numerous attempts to sink her. The problem was basically one of topography, for it was almost suicidal to make a low-level bombing run against a ship moored close to the side of a fjord. On the solitary occasion that a 750-pound bomb hit her it penetrated all the decks but failed to burst, but the X-Craft attack in September 1943 damaged her machinery and turrets severely. This forced the Germans to move her to a more southerly base for repairs, and once at Kaafjord she was within reach of RAF bombers flying from the northern USSR. On 15 September 1944 a force of 27 Lancasters

attacked with new weapons, the 12,000-pound Tallboy bomb and 400-pound JWII buoyant bombs. One hit forward caused great damage to the forepart of the ship and two near misses caused severe whip in the hull which put the machinery out of action once again.

Time was running out for the Third Reich and the high command decided that the best use for the *Tirpitz* was as a floating fortress or *schwimmende batterie* to defend Tromsö against invasion. With her damaged bow temporarily repaired the *Tirpitz* was moved from Kaafjord to Haakoy Island, three miles west of Tromsö in mid-October. Here she was at last within range of Lancasters flying from the far north of Scotland, and after an unsuccessful attempt on 29 October 32 Lancasters attacked her with Tallboys. For once the aircraft had it all their own way, evading radar cover until they were only 75 miles away and not meeting any defending fighters. In clear weather the bomb-aimers were able to achieve what they had been planning for three years and three hits were observed. Although still firing her antiaircraft guns the *Tirpitz* began to list heavily and 10 minutes after the attack had begun she rolled over and sank. It had taken 13 air attacks by 600 aircraft and an attack by midget submarines to finish her off, the last German capital ship.

The final duty for battleships in the European theatre was to provide covering fire for the big amphibious landings, starting with Torch and ending at Normandy. In June 1944 seven US and British battleships provided fire support for the D-Day landings, blasting concrete emplacements on the Atlantic Wall and disrupting German movements in response to calls from troops ashore. It was the last moment of glory for the older ships such as the USS *Arkansas* and *Nevada* and HMS *Warspite*, no longer able to take their place in the battle line but still capable of accurate shooting. They were only part of the massive effort which went into the liberation of Europe but they symbolized the way in which sea power had been the final deciding factor.

Above: The battleships of the *Littorio* Class were fast and elegant ships. This is *Vittorio Veneto* pictured on her trials.

9. END OF A REIGN

War in the Pacific had been brewing for many years, for as Japan grew in military and industrial power she grew more aware of just how flimsy the Western colonial empires were. The overriding need for Japan was raw materials like tin, rubber and oil to sustain her economic growth. Apart from coal the Japanese Home Islands lacked most of the important raw materials, and as these existed in abundance in the East Indies it was inevitable that Japanese thoughts should turn to conquest.

Standing in the way of expansion was the United States, and in particular her Pacific Fleet based on Pearl Harbor. With the British fully engaged in fighting the Germans and Italians they would be unable to spare reinforcements but the US Navy would have to be neutralized before any action could be taken in the East Indies. By 1940 the plans were ready for a daring strike to knock out the American Fleet and then occupy a huge defensive perimeter of island bases across the Pacific. Behind this island barrier the Japanese hoped to be able to absorb any counter attack and reduce enemy strength by attrition from submarines. The British attack on Taranto, needless to say, was studied with great interest for the Japanese intended to repeat it on a much larger scale, using far more modern aircraft in greater numbers.

At the end of November 1941 a fleet of six aircraft carriers, two battleships and three heavy cruisers put to sea for the attack on Pearl Harbor. Avoiding shipping routes the task force was in position 375 miles north of Hawaii by the night of 6–7 December without being detected. At 0700 hours next morning the first wave of aircraft left the carriers, and only two hours later the raid was over. The impossible had happened; the great base had been taken

by surprise and eight battleships had been sunk or badly damaged. Most of the damage had been done by the first wave, for the aircraft were able to identify Battleship Row easily. At 0810 hours a hit by a 1600-pound bomb on the *Arizona* detonated her forward magazines and she was blown apart. The *Oklahoma* capsized and the *California*, *Maryland*, *Tennessee* and *West Virginia* were badly damaged. The *Nevada* was set on fire and was nearly sunk in the harbor entrance but managed to put herself ashore leaving the fairway clear. The only battleship to escape serious damage was the *Pennsylvania*, but even she required extensive repairs and to all intents and purposes the US Pacific Fleet had been neutralized.

There were a few compensations. The American carriers had all been away exercising on the fatal Sunday and in their enthusiasm the Japanese pilots had omitted to destroy the huge tank farms. If the 4.5 million barrels of oil fuel stored there had been destroyed, Pearl Harbor would have been finished as a base, no matter how many ships had survived.

The next to feel the weight of Japanese air power were the British, who had finally sent the *Prince of Wales* and the *Repulse* to the Far East at the eleventh hour in the hope that such a gesture might intimidate the Japanese. The attack on Pearl Harbor robbed these ships of any strategic value but in any case they were doomed for there was only rudimentary air cover from a few obsolescent RAF fighters. Three days after Pearl Harbor as the two ships were searching for a reported amphibious landing on the east coast of Malaya they were attacked by a mixed force of 30 high-level bombers and 50 torpedo-bombers. The *Repulse* was slightly damaged by a bomb but the *Prince of Wales* was hit aft by a torpedo. This caused an extraordinary amount of damage by warping the port outer propeller shaft. Because the engine-room staff were not quick enough in disconnecting the turbine the shaft kept revolving and opened up all the after bulkheads. Within minutes the ship had taken on some 2500 tons of water in the machinery compartments and was listing 11.5 degrees. Then the shock of near-misses knocked out all the electric generators and the antiaircraft mountings lost power. A second wave of attacks missed both ships but the third wave put four more torpedoes into the starboard side of the battleship. By now the *Prince of Wales* was doomed for her pumps could not cope with the progressive flooding although she could still steam. She made off slowly to the north and survived a further bomb hit at 1244 hours but at 1320 hours she suddenly lurched further to port and capsized, taking with her Admiral

Previous page: The USS *New Mexico* hit by a kamikaze, one of two occasions in 1945 when she suffered damage. The mast and derricks on the left belong to a transport.

Below: AA gunners on the fantail of the *Massachusetts* take advantage of a lull in operations.

Phillips, Captain Leach and 327 men.

The *Repulse*, despite her age, showed great skill in dodging the attacks but the third attack mentioned above scored a hit aft. Unable to steer, she was helpless and took three more torpedoes. Captain Tennant ordered Abandon Ship and destroyers were able to rescue him and 796 of the crew. Following so soon on the heels of Pearl Harbor it was the end of nearly a century of Western supremacy in the Far East. It also marked the end of the battleship's supremacy for even if it could be argued that none of the other ships was modern, the *Prince of Wales* was a new-generation ship, designed specifically to survive air attacks. For many years the principal cause of her loss was widely believed to be bomb-hits, and not until the wreck was examined in 1966 by skindivers was it realized that no bombs had penetrated her decks. The real cause was the lack of air cover; only four years later four sisters of the *Prince of Wales* operated with impunity off the Japanese Home Islands.

In retrospect the destruction of the battle fleet at Pearl Harbor was a blessing for it freed the American naval aviators from pre-war concepts of operating carriers as an adjunct to the battle squadrons. Now the fast carrier task group had to be the main striking force because there was no other, and the first big battles in 1942, Coral Sea and Midway, were decided by rival air groups without the surface fleets making contact at all. Particularly at Midway, when the Japanese Commander in Chief, Admiral Yamamoto, had seven battleships, including the giant *Yamato* and yet was powerless to defeat three American carriers. In 1942 both sides finally concluded that the battleship was no longer relevant and all design work on battleships was suspended in favor of greatly expanded carrier programs. By this time the American 1937 ships were in service and the *South Dakota* Class were nearly ready but these were not the only American battleships being built. Even before the war the USN realized the importance of battleships in the carrier escort role and since the new *Essex* Class carriers were to be 33-knot ships it was felt that

the new battleships should have the same maximum speed.

Freed from the treaty restrictions the US Navy could build the ships that it wanted, and in 1939–40 six 45,000-ton *Iowa* Class were authorized, followed by five *Montana* Class displacing 56,000 tons. The *Iowa*s were magnificent ships, fast and long-legged to keep up with the carriers, whereas the *Montana* Class had heavier armor and three more 16-inch guns but six knots less speed as they were intended for the battle line. The *Montana*s and two of the *Iowa*s were stopped in 1943 because of steel shortages and the four *Iowa*s would not be ready until late 1943. In the meantime the burden had to be borne by the new battleships and the veterans. With a tremendous effort the Pearl Harbor casualties were repaired, apart from the *Arizona* and *Oklahoma*, some to be sent back into service with only updated antiaircraft batteries and others after total rebuilding.

Above: *Colorado* in 1942. She was refitting on the West Coast at the time of Pearl Harbor and had to wait until 1944 for modernization.

Below: The *Hyuga* running trials after completion of her massive reconstruction in September 1936.

Right: The giant *Yamato* fitting out at Kure in September 1941. The face armor of the 18-inch turret in the foreground was 25.5 inches thick.

Below: The reconstructed *Nagato* lying at Tsingtao during the Sino–Japanese war.

The Japanese for their part decided to convert the incomplete *Shinano* into a giant aircraft carrier, not a fleet carrier in the accepted sense but a ship capable of repairing and rearming other carriers' aircraft. They also removed the after 14-inch turrets from the *Hyuga* and *Ise* to provide space for a hangar, flight deck and catapults for launching float-plane bombers. It was a cumbersome conversion for the midships turrets were badly masked by the new superstructure, and by the autumn of 1943 when they recommissioned neither the aircraft nor the pilots were available.

The Pacific War was not fought entirely between carrier air groups, and when US forces landed on Guadalcanal at the end of 1942 powerful Japanese surface forces were thrown in to try to dislodge them. On the night of 11–12 November 1942 the fast battleships *Hiei* and *Kirishima* attempted to bombard the US Marines' airstrip at Henderson Field but were surprised by a force of US cruisers and destroyers. In a fierce short-range action the *Portland* and *San Francisco* inflicted severe damage on the *Hiei*, and after she took a probable torpedo-hit she withdrew to the north. Her sister *Kirishima* escaped with only one 8-inch hit but planes from the carrier *Enterprise* found *Hiei* next day and harried her mercilessly. Finally after 300 men had died in fires and explosions she was abandoned and sunk by her escorting destroyers.

The following night another bombardment was attempted but this time the battleships *Washington* and *South Dakota* were in support. At 2316 hours both battleships opened fire on a Japanese light cruiser, only to suffer the sort of unpleasant surprise the British had suffered

at Jutland. The Japanese were experts at night-fighting and they opened fire immediately with guns and torpedoes. All four American destroyers were put out of action before they could fire their own torpedoes, and to add to the confusion the *South Dakota* suddenly went out of control. About 17 minutes after the action had started the concussion from one of her twin 5-inch gun mountings caused a short-circuit in the electrical system. Although the power-loss lasted only three minutes the entire ship was in darkness and there was no power for guns, gyros or fire control. She turned to avoid the blazing destroyers but blundered off toward the Japanese line. At a range of only 5800 yards she was silhouetted against the glow and was fired on by the *Kirishima* and the heavy cruisers *Atago* and *Takao*.

The *South Dakota* was saved from serious damage because the *Washington* had prudently kept her searchlights switched off. Firing on radar the *Washington* was able to close to 8400 yards before riddling the *Kirishima* with nine 16-inch hits. The Americans withdrew to lick their wounds, leaving the Japanese destroyers to save the survivors of the *Kirishima* before sinking her with torpedoes. Although tactically an American victory it had revealed serious weaknesses in their organization. Without the advantage of radar they would certainly have beaten by the Japanese and that might have led to the loss of Guadalcanal.

As the new battleships joined the Pacific Fleet the older ships were relegated to bombardment duties in support of the various amphibious landings. The role of the fast battleships was not merely to ward off possible surface attacks on the carriers but to provide additional antiaircraft firepower. Being designed as steady gun-platforms their fire control was better than a cruiser's or a destroyer's, so that their effectiveness 'per

barrel' was greater. In the Battle of the Philippine Sea in June 1944 Admiral Willis A Lee's battleships, the *Washington, North Carolina, Iowa, New Jersey, South Dakota, Alabama* and *Indiana* were disposed in a battle line 15 miles east of the nearest carriers. Their task was to put up a wall of fire to thin out any Japanese carrier planes trying to attack Admiral Spruance's forces, and this they did superbly, inflicting many of the casualties suffered by the inexperienced Japanese pilots.

The last battle of the Pacific was also the biggest in history. Leyte Gulf, in reality four separate actions, saw all types of warship functioning as designed: battleships, cruisers and destroyers were all involved in possibly the last conventional surface actions which will ever be fought. The battle came about because the Japanese, in spite of the hammering that they had taken since the heady days after Pearl Harbor, still hankered after an annihilating battle between the two fleets. The demolition of their island perimeter and dwindling oil reserves made it imperative to do something decisive, and the invasion of the Marianas in mid-1944 finally forced their hand. It did not take a genius to predict that the next American thrust would be to the Philippines, and to defeat this the *Sho-1* or Victory Plan was conceived. It was to be a final gambler's throw with the entire surface fleet committed to tempt the Americans into full-scale action. It was crude but simple; without sufficient carrier aircrews to fly from the carriers the surface forces would have to force their way through to the invasion areas. Faced with this threat the Americans would have to bring in their main fleet and give battle.

Command of the Mobile Force was entrusted to Vice-Admiral Ozawa and it comprised four carriers and the two hybrid battleship-carriers *Hyuga* and *Ise* (but without aircraft) as well as three cruisers. Vice-Admiral Kurita had Forces A and B, the heavy striking force, comprising the *Yamato, Musashi, Nagato, Haruna* and *Kongo* and 12 cruisers. Force C was divided into a Van Squadron under Vice-Admiral Nishimura, with *Fuso* and *Yamashiro* and a single heavy cruiser, and a Rear Squadron under Vice-Admiral Shima, with three heavy cruisers. To the Americans these formations were identified by their location: the Mobile Force was labelled the Northern Force, Forces A and B became the Center Force and Force C the Southern Force.

Ozawa's role was that of decoy to draw Admiral Halsey's Fast Carrier Task Force away from the invasion fleet. Forces A and B would then join Force C to destroy the invasion fleet, brushing aside any opposition. Against

the 18-inch guns of the *Yamato* and *Musashi* and the fearsome Long Lance torpedoes of the cruisers and destroyers would be ranged only six old battleships, the *Mississippi*, *Maryland*, *West Virginia*, *Tennessee*, *California* and *Pennsylvania* for the six fast battleships were with Halsey's Third Fleet. Decoying Halsey would also reduce the threat from air attack, it was hoped, for the attackers would only be faced by the Seventh Fleet's escort carriers. It was accepted that Ozawa's force would probably be destroyed but, faced with the risk of certain defeat within a few months, the Japanese felt that the sacrifice would be worthwhile.

As soon as the first assault waves were reported moving into Leyte Gulf *Sho-1* was put into action, Ozawa sailing from Japan and Kurita, Nishimura and Shima from Brunei. But things went wrong almost immediately for Kurita's heavy units were sighted by two US submarines as they passed through the Palawan Passage. After sending off the vital news the two submarines made a brilliant attack, torpedoing three heavy cruisers. The defending air forces in the Philippines had wasted their strength in largely unsuccessful attacks on American carriers rather than provide a combat air patrol over Kurita's ships and so

they now felt the full weight of air attack. Over 250 planes from Task Force 38 mounted five separate strikes.

The *Yamato* and the *Nagato* were each damaged by two bomb hits but the *Musashi* bore the brunt. An estimated 13 torpedoes hit her on the port side and seven more on the starboard side, as well as 17 bombs and 18 near-misses. Even her massive protection could not stand up to such punishment and Kurita was forced to leave her behind. She finally sank about eight hours after the attacks had begun, but the fifth attack was the last. Kurita was not to know it, but Halsey had taken the bait and the whole of his Fast Carrier Task Force was in hot pursuit of Ozawa, leaving the invasion armada off Samar undefended. To the appalled Americans it was a nightmare come to life, the giant *Yamato* and her consorts attacking flimsy escort carriers and their destroyer escorts. Nor were the CVEs' aircraft armed with weapons for attacking battleships; their job was to provide support for the troops ashore, and no dive-bombers were embarked. And yet the impossible happened – the Japanese withdrew without achieving the destruction of the invasion fleet, although they sank the escort carrier, the *Gambier Bay*, two

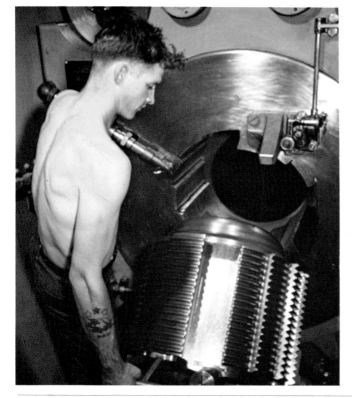

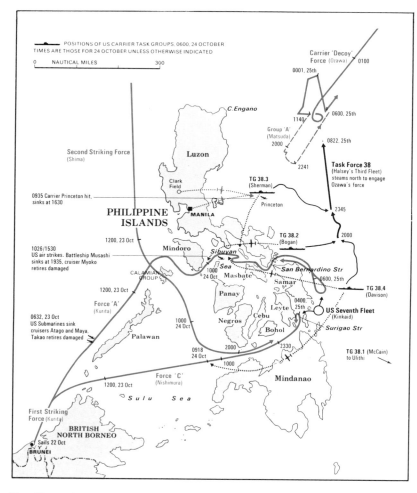

POSITIONS OF US CARRIER TASK GROUPS. 0600. 24 OCTOBER
TIMES ARE THOSE FOR 24 OCTOBER UNLESS OTHERWISE INDICATED

0 NAUTICAL MILES 300

Carrier 'Decoy' Force (Ozawa) 0100
0001, 25th
0600, 25th
1140
Group 'A' (Matsuda) 2000
0822, 25th
2241
Task Force 38 (Halsey's Third Fleet) steams north to engage Ozawa's force

Second Striking Force (Shima)
Luzon
Clark Field
TG 38.3 (Sherman)
2345
0935 Carrier Princeton hit, sinks at 1630
PHILIPPINE ISLANDS
MANILA
Princeton
2000
TG 38.2 (Bogan)
1200, 23 Oct
Mindoro
Sibuyan
1000 24 Oct
San Bernardino Str
0600, 25th
1026/1530 US air strikes. Battleship Musashi sinks at 1935, cruiser Myoko retires damaged
Sea
Masbate
Samar
0400, 25th
TG 38.4 (Davison)
CALAMIAN GROUP
1200, 23 Oct
Panay
Leyte
US Seventh Fleet (Kinkaid)
Force 'A' (Kurita)
Negros
Cebu
Bohol
Surigao Str
0632, 23 Oct US Submarines sink cruisers Atago and Maya. Takao retires damaged
1000 24 Oct
1000 24 Oct
2000
2330
TG 38.1 (McCain) to Ulithi
Palawan
0918 24 Oct
1000
Mindanao
1200, 23 Oct
Force 'C' (Nishimura)
S u l u S e a
First Striking Force (Kurita)
BRITISH NORTH BORNEO
Sails 22 Oct
BRUNEI

Top: The newly reconstructed *Yamashiro* running trials over the measured mile in Tateyama Bay, December 1934.

Right: The *Colorado* and *Texas* surrounded by a giant armada of transports, LSTs and smaller warships in Leyte Gulf, October 1944.

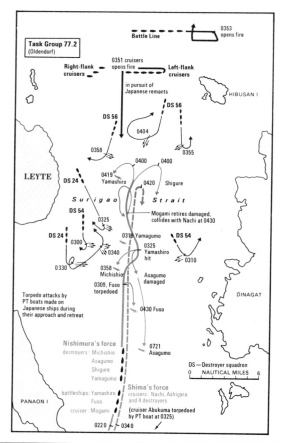

Task Group 77.2 (Oldendorf)

0353 opens fire
Battle Line
0351 cruisers opens fire
Right-flank cruisers
Left-flank cruisers
in pursuit of Japanese remnants
HIBUSAN I
DS 56
DS 56
0404
LEYTE
0358
0355
0400 0400
DS 24
0419 Yamashiro
0420 Shigure
S u r i g a o   S t r a i t
DS 54
0325
Mogami retires damaged, collides with Nachi at 0430
DS 24
0319 Yamagumo
DS 54
0300
0325 Yamashiro hit
0310
0340
0330
0358 Michishio
Asagumo damaged
0309, Fuso torpedoed
DINAGAT
Torpedo attacks by PT boats made on Japanese ships during their approach and retreat
0430 Fuso
0721 Asagumo
Nishimura's force
destroyers: Michishio
Asagumo
Shigure
Yamagumo
battleships: Yamashiro
Fuso
cruiser: Mogami
DS = Destroyer squadron
0 NAUTICAL MILES 6
Shima's force
cruisers: Nachi, Ashigara
and 4 destroyers
(cruiser Abukuma torpedoed by PT boat at 0325)
PANAON I
0220   0340

destroyers and a destroyer escort.

Historians have wrangled ever since over the reason for Kurita's sudden withdrawal when victory was within his grasp. From the Japanese point of view things looked very different. For one thing the admiral had already lost his cruiser flagship, he was recovering from fever and was not a young man. All day his squadron had fought off attacks by submarines and aircraft, losing four heavy cruisers and the *Musashi*. When he found the sea off Samar empty of ships and aircraft he was a worried man, for he could hardly believe that Halsey had swallowed the bait so completely. The escort carriers' planes were able to land ashore, refuel and rearm and then return to the attack, so that the Japanese had no idea of just how many planes there were. The heroic efforts of Admiral Sprague's pilots concealed the fact that there were no dive-bombers and eventually, one suspects, the tension proved too much for Kurita and he withdrew.

Part of Kurita's problem was that he suspected some disaster had befallen Force C, which was heading through Surigao Strait, but he did not know for certain. Admiral Nishimura was looking forward to a night action for he was certain that the *Fuso* and *Yamashiro* would be the equal of the Americans. Nor did Kurita know for certain that Halsey had been tempted by Ozawa's decoy force. Given the tremendous strain the Japanese force had been under ever since leaving Brunei it can be understood how the frustration of the fighting off Samar finally induced Kurita to give up.

As predicted Ozawa's force was devastated when Halsey's planes caught up with it off Cape Engaño. All the carriers were sunk but the carrier-battleships *Hyuga* and *Ise* both escaped and made their way back to Japan. But the most devastating defeat of all had overtaken Nishimura when he entered Surigao Strait just after midnight on 24–25 October. His force, comprising two destroyers leading the flagship *Yamashiro*, the *Fuso* and the heavy cruiser *Mogami* in line ahead with two more destroyers guarding the flanks, brushed aside attacks by PT-Boats and was then attacked by American destroyers. Nishimura appears to

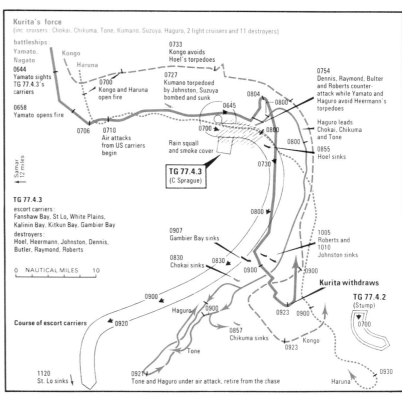

Kurita's force
(inc cruisers: Chokai, Chikuma, Tone, Kumano, Suzuya, Haguro, 2 light cruisers and 11 destroyers)

battleships:
Yamato,
Nagato
0644
Yamato sights
TG 77.4.3's
carriers

Kongo
Haruna

0733
Kongo avoids
Hoel's torpedoes

0700
Kongo and Haruna
open fire

0727
Kumano torpedoed
by Johnston, Suzuya
bombed and sunk

0804
0800

0754
Dennis, Raymond, Bulter
and Roberts counter-
attack while Yamato and
Haguro avoid Heermann's
torpedoes

0658
Yamato opens fire

0645

0700

0800

Haguro leads
Chokai, Chikuma
and Tone

0706

0710
Air attacks
from US carriers
begin

Rain squall
and smoke cover

0800

0730

0855
Hoel sinks

TG 77.4.3
(C Sprague)

0800

TG 77.4.3
escort carriers:
Fanshaw Bay, St Lo, White Plains,
Kalinin Bay, Kitkun Bay, Gambier Bay
destroyers:
Hoel, Heermann, Johnston, Dennis,
Butler, Raymond, Roberts

0907
Gambier Bay sinks

1005
Roberts and
1010
Johnston sinks

0830
Chokai sinks

0830

0900

0900

Kurita withdraws

TG 77.4.2
(Stump)

0920

0900

Haguro

0900

0923

0900

0700

Course of escort carriers

0857
Chikuma sinks

Tone

0923

Kongo

1120
St. Lo sinks

0921

Tone and Haguro under air attack, retire from the chase

0930

Haruna

0 NAUTICAL MILES 10

← Samar 12 miles

have taken no evasive action, and at 0307 hours a spread of probably five torpedoes hit the *Fuso* amidships. Oil fuel caught fire and then a series of explosions tore the ship in half, but instead of sinking the two burning halves drifted apart. Both Japanese and American lookouts reported two blazing ships, and the after section took an hour to sink.

Behind the destroyers and PT-Boats was waiting Admiral Jesse B Oldendorf's Battle Line, old battleships but equipped with the latest fire control and radar. At 0353 hours they opened fire at 22,800 yards, first the *Tennessee* and *West Virginia* and then the

Top: The *Kirishima*, a
sister of the *Kongo* in
Sukomo Bay in 1937.

Above: The *Yamato*
running trials in October
1941.

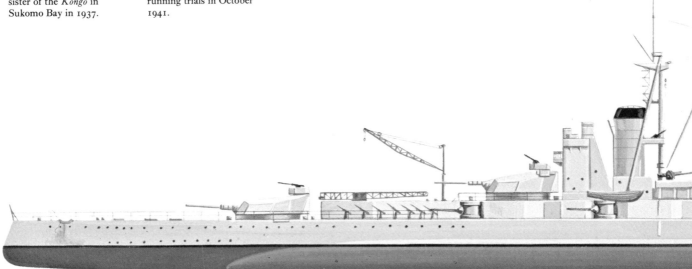

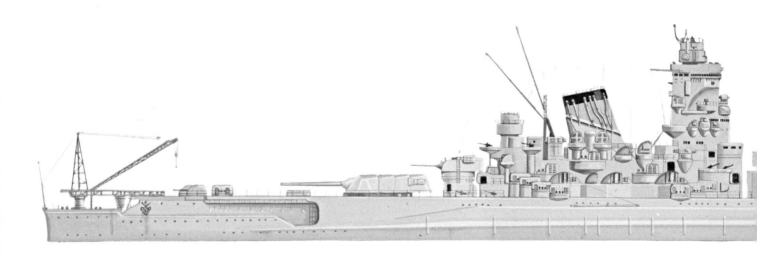

Above: Official model of the American reply to the *Yamato*, the 56,000-ton *Montana* Class. They were to be armed with twelve 16-inch guns and protected by 16-inch belt armor.

Left: The old battlecruiser *Kongo* was rebuilt in 1936–37 as a high-speed battleship to enable her to escort carriers. This involved lengthening the hull and installing much more powerful machinery.

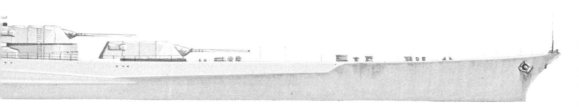

Left: The *Yamato* was the ultimate in battleships, with nine 18-inch guns and the heaviest armor ever made. Yet she and her sister *Musashi* never justified their enormous cost.

Maryland and the flagship *Mississippi*. Incredibly the *Yamashiro* seemed impervious to broadsides of 14-inch and 16-inch shells, even when hit by more torpedoes, but finally she slowed to a dead stop and lay blazing furiously in the water. No ship could take that sort of pounding indefinitely and at 0419 hours she finally rolled over and sank. Surigao Strait was a fitting swansong for the battleship, particularly as both the Japanese and the American ships were veterans of an earlier generation. They may have been overtaken by the carriers in importance but when it came down to a question of stopping a strong force of ships, just as the British had found with the *Bismarck*, the battleship's guns were the final arbiter.

The Imperial Japanese Navy was all but wiped out at Leyte, for there were aircraft but no trained pilots to fly them and no carriers; there were still surface ships but no fuel to enable them to put to sea. As the remnants of the air force immolated themselves in *kamikaze* attacks on the invasion fleet around Okinawa the navy planned the biggest suicide mission of all. The giant *Yamato* was ordered to use the remaining oil fuel (there was only enough for a one-way trip) for a last sortie against the invaders. Although there was talk of her blasting her way through the ring of Allied ships and then beaching herself on Okinawa as a huge gun emplacement, the real purpose was to act as live bait. By drawing off as many carrier planes as possible it was hoped to leave the air-space around Okinawa free for a gigantic *kamikaze* attack on the transports. Codenamed *Ten-Go*, the force comprised the *Yamato*, the light cruiser *Yahagi* and eight destroyers under the command of Vice-Admiral Ito.

At 1600 hours on 6 April 1945 the *Ten-Go* force slipped away from Tokuyama Bay and headed toward Okinawa in a ring formation with the *Yamato* in the centre. At 1220 hours the next day the *Yamato* signalled that she could see large numbers of aircraft 33,000 yards off her port bow. At 1232 hours she opened fire, using even the 18-inch guns to fire a splash barrage against low-flying attackers. At 1240 hours the first bombs hit her and 10 minutes later she was hit on the port side by torpedoes. After another eight torpedoes on the port side and two on the starboard side the flooding got out of control and the list could no longer be corrected. After the last torpedo hit at 1417 hours the giant ship was listing 20 degrees and the order to abandon ship was given. She finally capsized and erupted in a huge explosion, probably caused by internal fires reaching the magazines. With her went 2498 officers and men, and the last of the invincible spirit of the Imperial Japanese Navy.

Although the fast carriers had dominated the Pacific War the battleship retained her prestige to the end. When General MacArthur and Admiral Nimitz witnessed the unconditional surrender of Japan it was on the quarterdeck of the USS *Missouri*, while HMS *Duke of York* was moored nearby representing the British Pacific Fleet. It has been claimed that battleships gave way to carriers because of their expense but this is not true; the carriers in Tokyo Bay in August 1945 were as costly to build and operate as any battleship. What had been superseded was the heavy gun, for its maximum range was only 20–25 miles whereas the carrier's planes could reach out 300 miles. As we have seen the battleship's heavy anti-aircraft firepower was invaluable to any carrier task force and she acted as a deterrent to any surface attack on the carriers.

The battleship had a hard war but she was by no means the sitting duck that people seem to think. As the table shows only one Allied unit was sunk at sea by a submarine and only

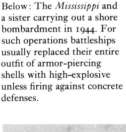

Below: The *Mississippi* and a sister carrying out a shore bombardment in 1944. For such operations battleships usually replaced their entire outfit of armor-piercing shells with high-explosive unless firing against concrete defenses.

Left: The *Washington* alongside the repair ship *Vestal* in February 1944, following a collision with the *Indiana*.

Left: The *Iowa* and a sister (probably the *New Jersey*) in company in 1944. Only these elegant ships had the speed and endurance to keep up with the *Essex* Class fast carriers.

Below: The *Idaho* bombarding Okinawa in April 1945 with a destroyer in support.

one modern battleship was sunk by air attack. Being expensive investments battleships were repaired whenever possible. The list of losses does not include all the Pearl Harbor and Taranto casualties, for example. Even after ships had been sunk their guns could fight on as coastal batteries, and in one case, the Soviet *Marat*, a turret was put back in working order although the ship was lying on the bottom of Kronstadt harbor.

It is difficult to choose the most distinguished battleship of World War II, for many of them performed heroic service of one form or another. The Japanese giants and the *Bismarck* are fascinating because of the colossal punishment they endured before sinking. Others like the *Warspite* bore a charmed life, surviving action after action without being sunk. She was bombed off Crete, was hit by two glider bombs

Above: The distinctive orange flame of cordite as a battleship (probably the *Idaho*) bombards Okinawa.

Below: The newly commissioned USS *Missouri*, the 'Mighty Mo' lies at anchor, 1944.

off Salerno and then touched off a magnetic mine in 1944, and still carried out a bombardment later before going to a well-earned retirement. Others like the *Prince of Wales* never shook off a reputation for bad luck. Other omens were noted too. Only after the sinking of the *Hood* was it remembered that she had been launched by Lady Hood, the widow of the admiral who lost his life when HMS *Invincible* blew up at Jutland. Although she saw no action apart from firing at enemy aircraft the American public held the 'Mighty Mo' in greater reverence than any other battleship. Even the biggest of ships had personalities as varied as the men who lived and fought in them.

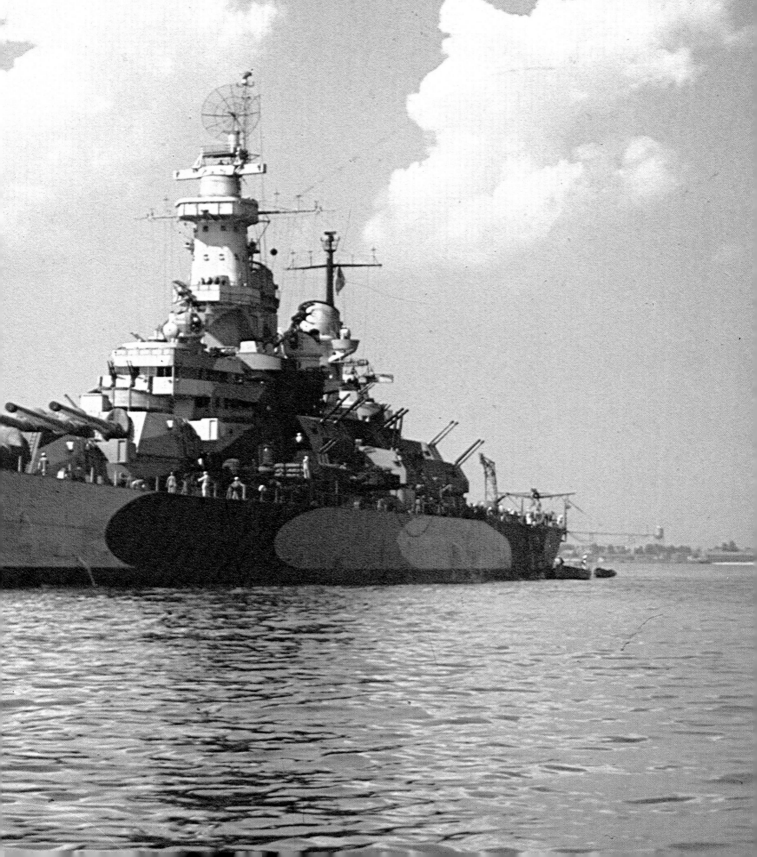

Battleship Losses 1939–1945

Name/Nationality	Date	Cause
United States		
Arizona	7 Dec 1941	Air attack, Pearl Harbor
Oklahoma	7 Dec 1941	Air attack, Pearl Harbor
Italy		
Conte di Cavour	12 Nov 1940	Incompletely repaired, sunk by air attack 1945
Roma	9 Sep 1943	German glider bombs, Mediterranean
Greece		
Kilkis	23 Apr 1941	Air attack, Piraeus
Lemnos	23 Apr 1941	Air attack, Piraeus
Soviet Union		
Marat	23 Sep 1941	Air attack, Kronstadt
Germany		
Admiral Graf Spee	17 Dec 1939	Scuttled Montevideo
Bismarck	27 May 1941	Gunfire and torpedoes, Atlantic
Scharnhorst	26 Dec 1943	Gunfire and torpedoes, Arctic
Tirpitz	12 Nov 1944	Air Attack, Norway
Gneisenau	28 Mar 1945	Scuttled Gdynia
Admiral Scheer	9 Apr 1945	Air attack, Kiel
Lützow	4 May 1945	Scuttled Swinemunde

Japan		
Hiei	13 Nov 1942	Air attack, Guadalcanal
Kirishima	15 Nov 1942	Gunfire and torpedoes, Guadalcanal
Mutsu	8 Jun 1943	Internal explosion, Japan
Musashi	24 Oct 1944	Air attack, Sibuyan Sea
Fuso	25 Oct 1944	Torpedoed, Surigao Strait
Yamashiro	25 Oct 1944	Gunfire and torpedoes, Surigao Strait
Kongo	21 Nov 1944	Torpedoed by submarine, Taiwan
Yamato	7 Apr 1945	Air attack, SW of Kyushu
Hyuga	24 Jul 1945	Air attack, Kure
Haruna	27 Jul 1945	Air attack, Kure
Ise	28 Jul 1945	Air attack, Kure
Great Britain		
Royal Oak	14 Oct 1939	Torpedoed Scapa Flow by *U-47*
Hood	24 May 1941	Blown up in *Bismarck* action
Barham	25 Nov 1941	Torpedoed in Mediterranean by *U-331*
Prince of Wales	10 Dec 1941	Air attack, Malaya
Repulse	10 Dec 1941	Air attack, Malaya
France		
Bretagne	3 Jul 1940	Gunfire, Mers-el-Kebir
Provence	27 Nov 1942	Scuttled Toulon
Dunkerque	27 Nov 1942	Scuttled Toulon
Strasbourg	27 Nov 1942	Scuttled Toulon
Courbet	9 Jun 1944	Scuttled as breakwater, Normandy

Below: Tired veterans like the *Colorado* had done considerably more work than the newer battleships.

10. THE GREAT SHIPS PASS

The battleship was officially dead but she would not lie down. A surprising number remained in commission after World War II, mainly because their large hulls provided useful accommodation for sailors and cadets under training. Their good communications equipment and spaciousness also made them ideal flagships.

The US Navy now had four of the magnificent *Iowa*s in commission, work having been stopped on the *Illinois* on 11 August, 1945 when she was only 22 percent complete. A sixth ship, *Kentucky*, was more advanced and she was permitted to go ahead on the understanding that she would complete in 1947. All the older battleships were rapidly decommissioned and laid up for disposal, with the exception of the *Mississippi*, which was earmarked for conversion to a gunnery training and trials ship. Even the new fast battleships, the two *North Carolina*s and the four *South*

*Dakota*s went into the 'mothball fleet' early in 1947, along with two enormous 'large cruisers,' the *Alaska* and the *Guam*, displacing a nominal 27,500 tons and armed with nine 12-inch guns. Although reminiscent of Fisher's Follies they were closer in concept to the Japanese 12-inch gunned armored cruisers built after Tsushima, for they had been built to outgun Japan's heavy cruisers rather than to masquerade as capital ships. Although a great combination of speed, endurance and gunpower they were hardly worth the time and resources spent on building them for they only appeared in mid-1944 and were very expensive to run.

The overriding question in 1946 was how the atom bomb would affect ships and naval warfare. On the one hand some pundits dismissed it as no more than a 'bigger bang,' while on the other hand the disciples of air power were predicting the end of all warships; the spirit of Billy Mitchell and Trenchard was

Previous page: A Sea Knight helicopter transfers supplies from the ammunition ship *Mount Katmai* to the *New Jersey* off Vietnam in July 1968.

Below: The *Iowa* cruising off the Korean coast in 1951.

rampant. To get a better idea of what the Bomb could do a series of tests was arranged for 1946 at Bikini Atoll in the Marshall Islands in the remote Pacific and large numbers of ships were allocated as targets. These included the veteran battleships *Arkansas*, *New York*, *Nevada* and *Pennsylvania* as well as old and damaged ships of other types from carriers down to landing craft. Also there was the battered *Nagato*, the only Japanese battleship to survive the last devastating carrier raids on Japanese harbors in July and August 1945.

The *Nevada* was moored at Zero Point, the planned dropping point for the first bomb,

surrounded by a cluster of some 20 ships of all types spaced at regular intervals within a two-mile radius. This formation bore no resemblance to wartime conditions, with ships much closer together than they would be in cruising formation, but it provided the best comparative data. Test A was planned for 1 July 1946 ('Able Day'), and a bomber was to be used to drop a plutonium bomb of the same type as that dropped on Nagasaki. A few seconds after 0900 hours observers saw a flash on the horizon followed by a brilliant fireball and then the familiar mushroom cloud appeared. Although the bomb had been intended

Above: The *New Jersey* in the North Atlantic in 1953. Apart from removal of catapults and floatplanes and all 20mm guns she is virtually unchanged from 1945.

Below: The *Iowa* with a carrier and a destroyer off Korea.

to burst directly over the *Nevada*, which had been fitted with a radio beacon and had bright orange upperworks to aid identification, it still missed the target by about 300 yards. She and the *Arkansas* and the heavy cruiser *Pensacola* were the only ships within half a mile of the explosion; their upperworks suffered badly but the hulls and turrets were virtually undamaged. The most notable effect on other ships was that many of them were radioactive.

The second test was carried out on 'Baker Day,' 25 July, and this time an underwater burst was planned, with the bomb suspended beneath the hull of a landing ship. The target ships were rearranged slightly, with none in the immediate vicinity of the explosion. This

time the explosion threw up a gigantic column of water at least 5000 feet high and a vast incandescent dome rose from the surface of the lagoon. For a moment the 26,000-ton *Arkansas* appeared to be lifted by the column of water and then she sank to the bottom of the lagoon. For several minutes the column, containing an estimated 1 million tons of water, continued to shed water and as this fell back it formed an expanding cloud of spray which engulfed half the 84 ships moored in the lagoon.

Once again radiation was a much greater hazard than blast. Although the water had absorbed the neutrons and gamma rays of the flash, masses of radioactive seawater washed

Far left: The *New Jersey* and the *Coral Sea* returning to the United States from Vietnam in April 1969.

Left: Sixteen years after the Korean War the thunder of 16-inch guns was heard again in 1968 when the *New Jersey* joined the 'gun line' off Vietnam.

Below left: A tug gently edges the *New Jersey* into dry dock as she prepares to go back into mothballs in October 1969.

Left: The British *Duke of York* comes alongside the quay at the end of her last voyage to the breakers in 1958.

Below: HMS *Duke of York* in her prime, serving with the Home Fleet in 1949.

over the decks of the ships and contaminated them. So complete was this contamination that it was impossible for some time to assess damage to machinery and equipment. Nearly eight hours after the blast the old carrier *Saratoga* succumbed to flooding and sank, while the *Nagato* was clearly in a bad way. Her bomb damage had never been repaired properly and makeshift repairs had been needed to get her to Bikini, but in spite of this she remained afloat for another two days before foundering quietly.

Although new warships would be built to incorporate the lessons of Bikini there was little point in worrying about existing ships. There were plans for finishing the *Kentucky* with an armament of missiles but they came to nothing as the cost was prohibitive, and in any case research on the sort of missiles suitable for use at sea was still at an early stage.

The cost of converting the existing ships was even higher for it would be necessary to remove triple 16-inch turrets, each one weighing as much as a destroyer, and much of the superstructure. In June 1959 a BuShips feasibility study showed that it might be possible to arm the *Iowa* Class with Twin Talos AA missile-launchers, two Tartar AA missile-launchers, an Asroc anti-submarine missile-launcher and four Regulus II cruise missiles. When this was

costed at $183,000,000 *per ship* BuShips came up with a cheaper conversion which left the forward 16-inch guns in and provided one Tartar, one Talos and one Asroc system and six Regulus II missiles. But even at $84,000,000 this was too expensive for ships which were already half-way through their effective lives.

An even more unrealistic scheme had been proposed for the *Washington* and *South Dakota* Classes in 1954 by the Chairman of the Ship Characteristics Board. This involved the removal of the after 16-inch turret and enlarging the machinery compartments to raise speed to 31 knots. To achieve such an increase it would have been necessary to raise shaft horsepower from 130,000 to 256,000 and to reconstruct the entire after hull form, and when the bill for this was estimated at $40,000,000 per ship, without allowing for any updating of the electronics and other equipment the subject was quietly dropped. The hull of the *Illinois* was demolished in its building dock starting in 1958 at the same time that the *Kentucky* was sold. The latter ship was much more complete than her sister, having been floated out of the dock in 1950, and before she was towed away for scrapping her boilers and turbines were removed and installed in two fast replenishment ships.

The Royal Navy had laid down a new

Below: Builder's model of the *Vanguard*, last and best of the long line of British battleships. She carried the most comprehensive outfit of AA guns, no fewer than 71 Bofors guns, most of which were in six-barrelled radar-controlled mountings.

battleship in 1941 and she was launched in 1944 as the *Vanguard*. The design had a chequered history for she had originally been conceived in 1938 as a fast battleship for the Far East, the reason being that all five *King George V*s and the *Lion* Class were earmarked for European waters, leaving only three modernized *Queen Elizabeth* Class and the *Nelson* and the *Rodney* to face the Japanese. At first it was proposed to build a fifth unit of the *Lion* Class, displacing 40,000 tons and armed with nine 16-inch guns but it was pointed out by the Director of Naval Ordnance that it would be impossible to provide any more guns and turrets before 1944 at the earliest. Then it was remembered that the 15-inch turrets from the *Glorious* and *Courageous* were still in store, and a new design was drawn up to use them instead. To save time the Engineer-in-Chief suggested cutting speed by a knot and duplicating the *Lion* machinery and although this was approved, late in 1939 plans had to be shelved for a while. When work stopped on the *Lion* Class it made sense to build a ship around the 15-inch guns to provide a good escort for carrier task groups and in 1941 approval was given to build her. It was hoped to get the *Vanguard* finished in time to see action in the Pacific in 1945 but last-minute efforts to improve her antiaircraft armament with the

latest fire control and multiple 40mm Bofors gun-mountings delayed completion until 1946.

It was fitting that the last British battleship should also be the best, and that the outstanding 15-inch gun should make a comeback. With a high flared forecastle the *Vanguard* was a magnificent seaboat, far better than any of her contemporaries. With a uniform antiaircraft armament of 71 40mm Bofors guns in a combination of 6- and 2- barrelled and single mountings controlled by nine separate fire control systems she also had the best anti-aircraft armament possible. Internally she was very similar to the *King George V* but with all the improvements suggested by war experience, particularly better auxiliary power arrangements and a higher degree of sub-division.

The *Lion* Class were now quite outmatched and it was intended to start two of them in 1946 to a completely recast design. No full details of this revised design have come to light but displacement would have risen to 56,500 tons with an increase of 100 feet in length and a massive increase in beam to 118 feet. A new 16-inch 45 caliber Mk IV gun was designed, with a higher rate of fire and better ballistics. Work on designing both guns and ships continued for some years but when it was realized that even 12-inch deck armor would

Below: HMS *Vanguard* firing a salvo from her 15-inch guns. Her flared forecastle helped make her a magnificent seaboat.

Above: The *New Jersey*
firing against the North
Korean shore. For service
off Vietnam it was necessary
to update her communi-
cations but, apart from
improving accomodation,
little more was done to her
in 1967–68.

Below: HMS *Anson* leaves
the Gare Loch on the Clyde
in 1958, bound for the
scrapyard. Two years later
the four *King George V*
Class would be followed by
the *Vanguard*.

Above: The Italian Navy
retained the *Andrea Doria*
and *Duilio* as training ships
until 1956, making them
two of the longest serving
battleships of the century.

not suffice against the latest bombs the project was abandoned. In 1948 the melancholy procession began as first the *Queen Elizabeth*s and *Royal Sovereign*s went to the scrapyard, with *Nelson* and *Rodney* bringing up the rear. As in the United States proposals for arming the *King George V* Class and the *Vanguard* with guided missiles were looked at but were rejected on grounds of cost. The *King George V*s served for a while on training duties but were all paid off into reserve by 1951, leaving the *Vanguard* to alternate between acting as Home Fleet Flagship and taking midshipmen to sea on training cruises.

France had lost the *Dunkerque* and *Strasbourg* when the order was given to scuttle the Toulon Fleet in 1942, to keep it out of German hands. The war had taken its toll of the 15-inch gunned battleships for both the *Richelieu* and *Jean Bart* had been damaged, the former by British action at Dakar and the latter by gunfire from the USS *Massachusetts* during the Torch landings in 1942. The *Richelieu* subsequently went to the United States for repair and served in the East Indies in 1944–45 but the *Jean Bart* needed major repairs before she went to sea again in 1949. She finally received her full secondary armament in 1955 and took part in the Suez operations the following year.

The Italians had chosen the right moment to bow out of the Second World War but this did not save the beautiful *Littorio* Class of which they were so proud. First the Germans unsportingly turned on their ex-allies, damaging the *Italia* and sinking the *Roma* with glider-bombs while they were trying to reach Malta. The surrendered fleet was the subject of much wrangling and to prevent either the *Italia* (the *Littorio* renamed in 1943 when Fascism went out of fashion) or the *Vittorio Veneto* from being claimed by the Soviets as reparations they were awarded to Great Britain and the United States. After lying idle in the Great Bitter Lake on the Suez Canal for some years they were both scrapped and the *Giulio Cesare* was given to the USSR. The Italians were finally allowed to retain the old *Andrea Doria* and *Duilio* for training purposes and the old ships lasted until the late 1950s before going for scrap.

The Soviet battleships had performed heroic feats as floating batteries during the Great Patriotic War, the *Marat* and *Oktybrskaya Revolutia* at Kronstadt and the *Parizhskaya Kommuna* at Sevastopol, but they saw little

Right: The *Missouri* in October 1950 firing at shore targets near Chong Jin, only 40 miles from the Chinese border.

Far right: Tugs struggle to free the *Missouri* after she ran aground in Chesapeake Bay in January 1950.

Below: The incomplete hull of the *Kentucky* being towed away for scrapping in October 1958. The bow section has already been used to repair the *Wisconsin* following a collision in 1956.

service at sea. The *Marat* was damaged beyond repair early in the siege of Leningrad but the other two (given back their original names *Gangut* and *Petropavlovsk* in 1942) survived several air attacks. They were too elderly to warrant more than running repairs after the war but survived as training ships until the late 1950s. The ex-Italian *Giulio Cesare* was a welcome addition to the Black Sea Fleet in 1948, despite her age and the fact that she had been scuttled once and sunk again during the war. As the *Novorossiisk* she was used for training for several years but in October 1955 she sank in Sevastopol when she set off an unswept wartime magnetic mine. Ironically she was sunk in almost the same place as the *Imperatritsa Maria* had blown up in 1916.

Although the 60,000-ton hulls of the *Sovyetskiy Soyuz* and *Sovyetskiy Byelorussia* and *Sovyetskaya Ukraina* had been abandoned and had been broken up in the late 1940s Western Intelligence sources continued to theorize about their existence. In the late 1940s rumours began to circulate about giant Soviet battle-

ships armed with 17-inch guns and modified versions of the German V2 rocket! Soon drawings were published in the Western press and it took many years for the myth to be dispelled. The story is now believed to have been an elaborate hoax concocted by an astute East Berliner, who sold bogus specifications to British Naval Intelligence.

The battleship might have remained in limbo, rather like a domesticated dinosaur, had it not been for the Korean War. When the North Korean Army crossed the 38th Parallel in 1950 the *Missouri* was the only US battleship in commission and in September that year she was sent to South Korea. Although she had been damaged by running aground in Chesapeake Bay the previous January she was able to serve three tours of duty between 1950 and 1953. She was joined by her three sisters and they proved invaluable for shore bombardment. Again and again the battleships were able to give rapid and precise fire support to ground troops, their biggest advantage being their ability to loiter and resume bombard-

ment if the enemy showed signs of further activity. No matter what weight of ordnance could be delivered by aircraft they always had to return to base after a short interval whereas the battleship was often still in the area and could be called up to provide more gunfire.

Although the last of the old battleships had been struck off the effective list in 1947 the *Tennessee*, *California*, *Colorado*, *Maryland* and *West Virginia* had to be kept in reserve to placate Congress and public opinion. Despite their age they were put back on the effective list at the time of the Korean War to prove that the US Navy was still up to strength, and there they remained until 1959 when permission to scrap them was finally given. The *Mississippi*, on the other hand, remained active from 1949 to 1956. Three of her triple 14-inch turrets were removed, leaving only the barbettes forward and aft and the aftermost triple turret. At first a series of 5-inch guns was mounted forward and a number of new radar sets were mounted in a special tripod mast. She was then given a twin 6-inch turret forward, similar to

the mountings in the latest cruisers, the *Roanoke* and *Worcester*. In 1952 she underwent a further conversion at Norfolk Navy Yard as the trials ship for the new Terrier antiaircraft missile. With two twin-arm launchers in place of the former 14-inch turret she carried out the evaluation of the first operational surface-to-air missiles. When she finally decommissioned at the end of 1956 she had been in almost continuous commission since 1917.

The four *Iowa*s remained active until 1957–58. In 1953 the *Iowa* and *Wisconsin* joined HMS *Vanguard* in the big NATO exercise Mariner in the North Atlantic, and showed off their enormous endurance by refuelling all the escorting destroyers in the task force. When the *Wisconsin* decommissioned on 8 March, 1958 at Bayonne, New Jersey it seemed that the end of the story had been reached, the first time since 1895 that the US Navy had no battleship in service. But soon the Vietnam War was absorbing an increasing American military effort and there was a vociferous demand from the US Marine Corps for something heavier than 8-inch gunfire support. Finally in 1967 permission was granted.

The choice fell on the *New Jersey*, last in commission a decade earlier. The *Missouri* was ruled out because her speed was limited after the bad grounding in Chesapeake Bay. The *Wisconsin* had also suffered damage; a fire during an overhaul had destroyed electrical circuits around the forward 16-inch turrets.

That left the *Iowa*, whose electronics were the least up-to-date, and so the decision was taken to cannibalise the *Iowa* and *Wisconsin* to speed up the refit of the *New Jersey*. She was taken in hand at Philadelphia Navy Yard in August 1967 and recommissioned the following April. During her refit the twin 3-inch/50 caliber AA guns were removed and the forward bridge was rebuilt to accommodate the latest communica-

tions and electronic warfare gear, and some improvements were made to habitability. The hardest problem, however, was to find sailors experienced in heavy gun turret drill and efforts were even made to find ex-Royal Navy men with similar experience. The *New Jersey*'s comeback was short but effective. On the gun line in Vietnam she spent 120 days in all, of which 47 days were continuous. During World War II she had fired a total of 771 16-inch shells, during the Korean War she had fired nearly 7000 rounds, but this time she fired 5688 rounds and 15,000 rounds of 5-inch as well. Sadly she was taken out of commission in 1969, for the navy discovered that it had no replacements for the liners of her 16-inch guns. A field full of spare liners was discovered just too late to reverse the process of redistributing her crew to other ships, but in the final analysis manning proved the biggest drawback to having a battleship back on the active strength. In the modern navy 70 officers and 1550 enlisted men are more precious than 16-inch shells. Yet despite the expense, in January 1981 Congress was asked once again to vote funds for reactivating the *New Jersey*.

Today there are still a surprisingly large number of battleships afloat preserved. The earliest to be saved was the USS *Texas*, acquired in 1948 for preservation as a memorial to the war dead of Texas. She rests in a permanent berth dredged into the San Jacinto battlefield, open to visitors. Three more states were able to acquire battleships as memorials, North Carolina, Alabama and Massachusetts. The *North Carolina* has been lying in the Cape Fear River near Wilmington since 1961, while the *Alabama* is exhibited near Mobile and the *Massachusetts* is berthed at Fall River, Massachusetts. The 'Mighty Mo' has a special place in American hearts as the scene of the Japanese surrender in 1945, and so she is earmarked for preservation as a national naval memorial. Although not the biggest battleships ever built the *Iowa* Class had the best balance of offensive and defensive qualities ever achieved and it is appropriate that one of them should remind future generations of the peak of perfection achieved.

Apart from wooden ships of the line, the oldest battleship in existence is HMS *Warrior*, the very first iron battleship. By a quirk of fate she survived as a training hulk until 1927 and was then towed to south Wales for service as a floating jetty at an oil terminal. For years navy oilers have berthed alongside to discharge oil across her deck and into storage tanks on shore. Below her massive gundeck survives intact, with the angled gunports still visible. She is to be preserved by Britain's Maritime Trust, on similar lines to the SS *Great Britain*, the world's first iron-hulled passenger ship. She has outlived all other British battleships for attempts to save first HMS *Warspite* in 1946 and then HMS *Vanguard* in 1960 came to nothing.

Below: The USS *Wisconsin* at sea just before the end of World War II, preparing to replenish fuel and stores. The *Wisconsin* was the last of the *Iowa* Class to be taken out of service in the 1950s.

Next in order of age is the only pre-dread-nought left, Admiral Togo's Tsushima flag-ship *Mikasa*. She was put in a permanent berth at Yokosuka dockyard after World War I but was nearly destroyed by air attacks in the closing weeks of World War II. At first the Occupation Forces wanted to demolish the wreck as a way of stamping out Japanese militarism but they were prevented by local officials who housed homeless refugees in the hull. The new Shogun, General MacArthur, showed a more tolerant attitude and subse-quently gave his wholehearted support to an appeal to raise funds for the *Mikasa*'s restora-tion. This was helped by the arrival of a Jutland veteran for scrapping in Japan. This was the Chilean *Almirante Latorre* which had fought at Jutland as HMS *Canada* and had then been

Below: The *New Jersey* on the 'gun line' in March 1969, near Tuyhoa. The forebridge has been built up and she has been fitted with a large number of communications and ECM antennae.

returned to the Chilean navy. Although a large dreadnought she had been British built and was able to provide a large number of fittings for the predreadnought *Mikasa*.

When we last head of the battlecruiser *Goeben/Yawuz Sultan e Selim* she was badly damaged by mines. Although a defeated partner of Germany, Turkey escaped the harsh terms meted out to Germany at Versailles, and under the Treaty of Sèvres was permitted to keep the *Yawuz Sultan e Selim*. The ship underwent a protracted repair and modernization at Izmir and with her name shortened to *Yawuz* (from 1936) she served another 40 years as the flagship of the Turkish navy. In 1966 she was offered for sale but in spite of strenuous efforts to save her she was finally sold for scrapping six years later. This was a particularly tragic loss for there was more than enough money available in Germany to preserve the last survivor of the High Seas Fleet even if there was none to spare in Turkey. But the political climate in Germany was hostile to any project which smacked of militarism.

An eyewitness who visited the *Yawuz* at Izmir reported that she was in remarkably good condition. In the turrets the original Krupp brass instruction plates with their Gothic lettering could still be seen and the sliding breeches were still in working order. Apart from losing her mainmast in 1941, when light AA guns were added aft, she was virtually unchanged since her commissioning in 1912. As a ship which had unwittingly changed the history of the world she had an overriding claim to be preserved but the time was wrong and she was too far from the limelight to attract publicity.

Today a fleet of battleships or a gun-duel between two giant ships is hard to imagine but the image conjured up by the word battleship is still powerful. The uninitiated label any big ship a battlewagon, and it is significant that when the Soviets completed a 22,000-ton missile cruiser recently the world's press hailed her as a battlecruiser. The *Kirov* bears no relation to any battleship ever built, either in appearance or function, but the description cruiser seems inadequate for a ship so much bigger than her predecessors.

Ships are still the biggest mobile structures on earth and to this status the battleship added another dimension. She had to stay afloat in the face of attacks from torpedoes, mines and shells and in addition stand the tremendous shock of firing her own ordnance and provide a home for her crew. Some achieved this compromise well, others failed, but all made their mark on naval history and we shall not see their like again.

INDEX/GENERAL

INDEX/SHIPS

ACKNOWLEDGEMENTS

The Author would like to thank the Agencies and individuals listed below for the use of their artwork and photographs:

M Aplin: pp 63 (center); 65 (top); 88 (top)
Author's Collection: pp 2–3; 87 (bottom); 113 (all four); 138–139; 146–147 (bottom); 153; 179
John Batchelor (artwork): pp 130–131 (all three); 142
Marius Bar: pp 15 (top); 34 (top)
Claus Bergen: p 105 (bottom)
Bibliothek für Zeitgeschichte, Stuttgart: pp 78–79 (bottom); 82 (below)
Bison: pp 37 (top); 40–41; 43 (top); 44 (center and top); 97 (top); 105 (top)
Bundesarchiv: pp 52 (top); 53 (center); 67 (top); 72–73 (bottom); 86 (top); 92–93; 94 (above); 106 (above)
Chas E Brown: p 119
Camera Press: p 150 (top)
Peter Endsleigh Castle (artwork): p 135
Central Naval Museum Leningrad via Boris Lemarchko: pp 32 (top); 42 (left)
Conway Maritime Press: pp 13 (top)
Conway Picture Library: pp 140–141 (bottom); 143; 177 (both); 180–181 (bottom)

Helen Downton (artwork): pp 164–165 (bottom two)
Photo Druppel: pp 56–57 (bottom); 73 (top); 78 (top); 83; 150
ECPA: p 13 (bottom); 84 (top)
Aldo Fraccoroli: pp 44–45 (bottom); 49 (bottom); 73 (center); 145 (both)
Robert Hunt: p 108–109
Imperial War Museum: pp 29 (bottom); 30 (top center and bottom); 33 (top); 34 (center); 35 (top); 36–37 (bottom); 43 (right); 46 (top left); 49 (top and center); 50; 53 (top); 57 (top); 63 (top); 65 (bottom); 66–67 (bottom); 68 (center); 69 (top); 70 bottom); 71 (top); 77 (all three); 79 (top); 82 (top); 83; 85 (center); 89 (bottom); 94–95 (bottom); 95 (above); 96 (above); 96–97 (bottom); 98–99 (main picture); 99 (above); 102–103; 106–107 (bottom); 107 (bottom); 108 (top two); 114–115; 118 (bottom); 123 (above); 129 (top); 136–137; 141 (top); 148–149 (bottom); 157 (bottom); 162–163 (top)
Humphrey Joel: pp 110; 111 (both)
Kriegsarchiv, Vienna: pp 32 (center); 68 (top two); 90–91 (bottom and top)
Photo Mayo: p 14–15 (bottom)
Musée de la Marine: pp 16 (all); 21 (bottom); 68 (center two); 72 (center); 84–85 (bottom);

85 (top); 128 (bottom)
Museo Storico Navale, Venice: pp 31 (top and center); 71 (bottom)
National Maritime: pp 8; 10–11 (both); 12; 28; 29 (top and center); 34 (bottom); 48; 54–55 (all); 61 (top); 62–63 (bottom); 66 (top); 72 (top); 76; 79 (center); 88 (below); 100–101; 122–123 (below); 152 (top)
National Archives: pp 1; 4–5
Richard Natkiel (maps): pp 45; 96; 98; 99; 103; 144; 146; 149; 151; 162; 163
Peter Newark Western Americana: p 19 (top)
Norwegian Naval Archives: pp 38–39 (bottom)
Officio Storico della Marina Militare, Rome: p 90 (top and center left)
Real Photographs: p 175 (top)
Science Museum, London: pp 9; 17; 178
Sjohistoriska Museat, Stockholm: p 18
Ufficio del Storico, Rome: p 70 (top)
US Air Force: p 125 (all three)
US Navy: pp 6–7; 19 (center and bottom); 20 (both); 21 (top); 22–23; 25; 26–27; 32–33 (bottom); 35 (bottom); 38 (top and center); 39 (top); 46 (top right and bottom); 47; 52 (bottom two); 53 (bottom); 58–59; 61 (bottom); 64; 74–75; 81; 87 (top and center); 98 (bottom); 104 (both);

112 (both); 116–117; 120–121; 126 (both); 127; 129; 132 (top); 134 (top); 154–155; 156; 157 (top); 158 (bottom); 160 (both); 161 (all three); 163 (bottom); 164 (top); 166; 167 (all three); 168–169 (both); 170–171; 172–173; 174 (above); 174–175 (bottom); 176 (both); 180 (top); 182–183; 184–185; 186–187; 188–189
US Navy via Shizuo Fukui: pp 158 (top); 164 (center)
P A Vicary: pp 42 (right); 51; 89 (top); 129 (center)
Alfredo Zennaro, Rome: pp 132–133 (bottom); 146 (center); 181 (top)

The author would like to thank Laurence Bradbury, the designer, Donald Sommerville, the editor, and Penny Murphy who compiled the index.